This book is due for return on or before the last date shown below.

# Key Ideas in Teaching Mathematics

Research-based guidance
for ages 9-19

Anne Watson
Keith Jones
Dave Pratt

# OXFORD
UNIVERSITY PRESS

Great Clarendon Street, Oxford, OX2 6DP,
United Kingdom

Oxford University Press is a department of the University of Oxford.
It furthers the University's objective of excellence in research, scholarship,
and education by publishing worldwide. Oxford is a registered trade mark of
Oxford University Press in the UK and in certain other countries

First published in 2013

Impression: 1

British Library Cataloguing in Publication Data

Data available

ISBN 978-0-19-966551-8

Printed and bound by
CPI Group (UK) Ltd, Croydon, CR0 4YY

# ACKNOWLEDGEMENTS

The preparation of this book and the associated web resource has been supported by many other people. First, substantial thanks go to Josh Hillman and the Nuffield Foundation who generously funded the work and are hosting the associated website. However, the views expressed in this book are those of the authors and not necessarily those of the Foundation. Second, we have been very well supported and encouraged by Keith Mansfield, Clare Charles, Victoria Mortimer, and Shanker Loganthan at Oxford University Press who helped us bring the book to fruition. Vinay Kathotia and Fran Bright at the Nuffield Foundation have worked enthusiastically to prepare the website and linkages to it. Thanks also to Yvonne Dixon who produced the index.

In writing the text, as well as helping each other we had the support of research assistants: Ellie Darlington, Phillip Kent, Willeke Rietdijk, Chris Wild, and Lizzie Kimber. We had an advisory panel of teachers and teacher educators who gave us some initial ideas about disseminating the work, and also read good drafts of all the chapters: Nicholas Andrews, Richard Cowley, Cosette Crisan, Linda Dongworth, Jackie Fairchild, Sarah Gilbert, Karen Russell, Sarah Shekleton, and Steve Shipman. Their wisdom and experience led us to produce a book which does not talk down to teachers, but is informative and gives them research knowledge with which to develop their work, while also pointing towards associated resources.

And finally, our partners, who suffer so much when we are writing late into the night but also provide wise and knowledgeable counsel: John Mason, Julie-Ann Edwards, and Christine Pratt.

# FOREWORD

This innovative and practical book brings together knowledge on how secondary school students best learn mathematics. The authors present the latest research evidence on teaching and learning, organised into seven key mathematical domains. The result is an accessible and engaging read that offers a comprehensive overview of secondary mathematics and the transitions into and out of it. It is designed to have a direct impact on teachers and others working in mathematics education.

The need for work in this area was identified in an earlier project funded by the Nuffield Foundation, *Key Understandings in Mathematics Learning*, published in 2009. Although the focus of this influential literature review was mathematics at primary level, Professor Anne Watson, the project's secondary education specialist, examined the available evidence on two secondary level topics: algebraic reasoning and modelling; and problem-solving and integrating concepts. In doing this, she identified the general lack of accessible research-based writing on teaching secondary mathematics. This book was conceived as a way to address that gap, and its research and execution has been a collaboration between three experts in the field.

Themes such as measurement, proportionality, and spatial perception permeate mathematics. Indeed, mathematics can be characterised as the connection and application of these themes. The authors make a strong and persuasive case for the 'key ideas' they have chosen to explore in this book. For each idea, the book synthesises the international research literature on how children develop their reasoning, understanding, and skills, and on related teaching approaches. Informed by the authors' own experience, it identifies areas where additional evidence is needed, and provides links to relevant examples and classroom activities. We believe it will be an invaluable resource for those who face the challenge of explaining these ideas and for helping learners to connect and apply them.

The Nuffield Foundation has long been associated with education research and development, and a current priority is to strengthen the link between research evidence and classroom practice, particularly in mathematics. We hope this book will go some way to doing that, by providing teachers and others with the relevant evidence to inform and adapt their work. In order to ensure the material in the book is as relevant and accessible to practitioners as possible, we worked with the authors to support a series of workshops for classroom teachers to ensure their practical needs and ideas informed the research design at an early stage.

We have also worked with the authors and Oxford University Press to develop a website to accompany the book. Although the book and the website serve as valuable resources independently of each other, we have supplied web links and QR (quick response) codes throughout the text to link to online activities that exemplify the ideas presented. The website can be found at www.nuffieldfoundation.org/key-ideas-teaching-mathematics. We hope this will be a useful addition for readers who wish to enhance their understanding through practical examples.

From the inception of the research project to the publication of this book, Anne Watson, Keith Jones, and Dave Pratt have demonstrated an unstinting commitment to fostering new models for bridging research and practice and ultimately, to supporting the mathematical development of learners. This is a commitment that the Foundation shares, and we are delighted to support this important book.

Josh Hillman
Director of Education, Nuffield Foundation

# CONTENTS

# ABBREVIATIONS

| | |
|---|---|
| ACME | Advisory Committee on Mathematics Education (UK) |
| ANOVA | Analysis of variance |
| ASE | Association for Science Education (UK) |
| BODMAS | Mnemonic for order of operations: Brackets, of, division, etc. |
| CAS | Computer Algebra System |
| CGI | Computer Generated Imagery |
| CSMS | Concepts in Secondary Mathematics and Science |
| dB | Decibel |
| DGS | Dynamic Geometry Software |
| EDA | Exploratory Data Analysis |
| ICT | Information and Communications Technology |
| IIR | Informal Inferential Reasoning |
| JMC | Joint Mathematical Council (UK) |
| KUML | Key Understandings in Mathematics Learning |
| MEA | Model Eliciting Activities |
| NCTM | National Council of Teachers of Mathematics (USA) |
| RPR | Ratio and Proportional Reasoning |
| SI | International System of Units |
| SOH-CAH-TOA | Mnemonic for Sine-Opposite-Hypotenuse etc. |
| SOLO | Structure of Observed Learning Outcomes |
| TIMSS | Third International Mathematics and Science Study |
| ZDM | Zentralblatt für Didaktik der Mathematik |

# CHAPTER 1

# Introduction to key ideas in teaching mathematics

## Introduction

In this chapter we introduce and explain the reasons why we have written this book, and how we wrote it. We provide, as an advanced organiser, some common themes that emerge throughout the book. In addition, we outline our methodology and talk about our perspective in relation to other suggestions of 'big ideas' and 'key ideas' in school mathematics. Finally, we provide an overview of the content of the book.

## The scope of this book

In 2008 the Nuffield Foundation commissioned and funded a synthesis of research about how children learn mathematics from 5 to 16. This review was undertaken by Professors Nunes, Bryant, and Watson and was launched in July 2009 as a collection of reports entitled 'Key Understandings in Mathematics Learning' (KUML). The process of reviewing research for the synthesis identified a dearth of accessible writing about how students learn the kinds of mathematics encountered beyond age 9 years. The scope of the KUML synthesis also prevented a detailed study across the breadth of the mathematics curriculum for these older students.

In the formal mathematics education of older students, the new ideas tend to be: clusters of concepts; new ways of working; new, and potentially confusing, representations; new formal ideas that are not readily experienced in everyday life. There are significant new concepts, for example, in probability, trigonometry, and functions. This book presents research about how students learn these and spans the growth of mathematical understanding between years 9 to 19. We take 9 to be an age when some, but not all, children begin to handle more abstract ideas of mathematics, but in most schools the content we cover would be more frequently found for ages 11 and above.

The research that focuses on aspects of mathematics for older students tends to involve teaching experiments promoting particular approaches, geared towards particular learning aims which are often as much social and motivational as intellectual. However, there are also some significant caches of work on teaching and learning specific topics, particularly those that tend to be taught instrumentally or procedurally at age 13 and beyond in response to their perceived difficulty, such as solving equations or proving theorems.

Another insight arising from the KUML study is that teachers at the older level do not have access to systematic knowledge about how children learn the elementary concepts that are drawn on at the later stage. Consequently the help they offer to students who need to revisit more elementary concepts is not always informed by research. During the work for KUML, and subsequently at workshops with teachers and teacher educators, it became apparent that there is a need for a further synthesis of research for teachers of older students, with a strong focus on how knowledge about learning can inform practice.

We have therefore produced this research-informed resource, which brings together, in focused sections, knowledge about mathematics, teaching and learning that is appropriate at this higher level. We have drawn on findings about early concepts and shown how they are relevant to the teaching and learning of secondary school mathematics, which in UK terms is usually from ages 11 to 18. We have built on this basis to incorporate research about teaching and learning at higher levels, and to illustrate throughout how formal and informal knowledge and reasoning interact.

## Synthesising research for this book

This book is a resource for teachers, teacher educators, textbook writers, and curriculum policy makers on mathematics at secondary level, focusing on the issues raised above. We have viewed mathematics as developing 'bottom-up' through learning rather than solely 'top-down' from an academic viewpoint. In every chapter we identify what young students might already know before they arrive in secondary school, either from previous schooling or from outside experiences, and then examine this in terms of the new ideas that have to be understood.

We have approached the task through a systematic synthesis of relevant research about conceptual growth, through education, in key areas of the secondary curriculum, drawing on:

- theoretical explanations regarding how children learn mathematics which have been supported by research;
- research about learning mathematics at secondary level in relevant subject domains;
- relevant summary reports of research in particular areas;
- research on students' errors and teaching experiments, where these also illuminate ways of learning concepts successfully;
- research about the advantages and disadvantages of particular teaching approaches.

We focus on research that is available in English and published in books (including international handbooks), book chapters, journals, and high-quality refereed conference proceedings. Occasionally we go beyond this scope for substantial bodies of research elsewhere. We have selected descriptive and experimental research, and both qualitative and quantitative studies. We have included research in which ICT tool use is the main focus in the context of mathematical learning, such as using spreadsheets to help students understand the concept of variable. We have included studies that show what is possible in certain special circumstances, such as learning about functions using graph-plotting software, or reasoning geometrically using dynamic geometry. We have used refereed studies of students' typical errors and methods where relevant because these help to identify the development of understanding. The bibliography for this book is necessarily lengthy, and provides access to the reader for further details and insights into all the topics we discuss in the book. If we had included all the sources we looked at it would have been double the length.

The usual expectation of a research synthesis is that it should focus on empirical studies and 'summarize past research by drawing overall conclusions from many separate investigations that address related or identical hypotheses'

(Cooper, 1998, p. 3). Such an approach, however, depends on having 'identical hypotheses'. This is impossible because the aims of mathematics education, in terms of what students should be able to do, understand, and apply, vary between countries, assessment methods, and even between teachers. For example, the learning experienced by students through an intervention that focuses on reproducing standard geometrical proofs will be different from that of students who are being taught to produce proofs for themselves – yet each group is learning about mathematical proof. Differences in the nature and content of mathematical learning, and the role of pedagogy, make it impossible to conduct critical experiments that have any meaning for real, varied, teachers working in real varied contexts with varied curriculum aims. Thus all research into secondary mathematics learning has to be understood in its own context, with its own goals and methods.

We have worked systematically within these constraints, and the book presents summaries of what is known from research about learning and teaching the key ideas we have identified. There is a huge amount of research in mathematics education, far too much to report in detail, so we have focused on research that can inform the teaching of mathematics in the ordinary schools with which we are familiar. In that respect, there is a degree of subjectivity in the process and it is entirely possible that different authors would have reported the research in an alternative way.

## Key ideas of school mathematics

We have organised most of the book around seven key mathematical domains. These are: relations between quantities and algebraic expressions; ratio and proportional reasoning; connecting measurement and decimals; spatial and geometrical reasoning; reasoning about data; reasoning about uncertainty; functional relations between variables. Chapter 9, the last chapter, focuses primarily on some advanced aspects of mathematics which are rooted in teaching and learning at a younger level but are developed much further in higher education.

Identification of key ideas in mathematics learning is a recurring issue for curriculum designers and policy-makers and different decisions are made for different purposes. For example, Siemon's work for very young children focuses on 'trusting the count ', place-value, multiplicative thinking, partitioning, and proportional reasoning (2011) as these provide a foundation for future elementary learning. At a more overarching level, a European Commission report takes a broad view and characterises 'Big Ideas' in mathematics as:

- having high potential for developing conceptual knowledge;
- having high relevance for building knowledge about mathematics as a science;
- supporting communication and mathematics-related arguments;
- encouraging reflection processes of teachers. (Kuntze et al., 2011, p. 8)

They include representations, proof, ideas of infinity, classifying, ordering, formalising as examples of these, but avoid giving definitive lists. Another approach is that taken by Schweiger (2006) who writes about 'fundamental ideas' as those which:

- recur in the historical development of mathematics (time dimension);
- recur in different areas of mathematics (horizontal dimension);
- recur at different levels (vertical dimension);
- are anchored in everyday activities (human dimension). (p. 68)

The reasons for our choice of seven come from several directions: identifying threads in the curriculum; identifying mathematical concepts that permeate mathematics; identifying mathematical concepts that seem to present students with difficulties; and identifying mathematical concepts that have strong implications for employment and citizenship. Our criteria for choice therefore include the last three of Schweiger's above, but include considerations of learning mathematics, as indicated by Kuntze et al. (2011). Focusing on older school students means we have not presented research on number and place-value, nor have we dwelt on formalising everyday understandings in elementary contexts, nor on developmental theories of learning. However, when we thought about students' learning, we could not avoid drilling down into younger children's experience because of the importance of building on existing knowledge. To make our final choice of chapters we used our knowledge as teacher educators and mathematicians as well as researchers and chose domains that indicated different kinds of reasoning and activity within mathematics for the age group under consideration, and related these closely to more typical curriculum areas.

Some key ideas in mathematics pervade the whole of the subject and are integrated into many chapters. This is true of proof, ICT use, and representation. In addition, some ideas that start in the elementary stages of mathematics education need to be elaborated and shown to be centrally relevant in the context of more complex mathematical ideas.

# Recurring themes in all chapters

We have identified a number of themes that seem to underpin considerations of learning mathematics irrespective of the content area. In Schweiger's terms (2006) these might be seen as horizontal and vertical fundamental ideas, with a long-lasting role in mathematics and a connection with how mathematics is done. In the terms of Kuntze *et al.* (2011) our themes would contribute to developing conceptual knowledge, to identifying and communicating what is mathematical, and also to pedagogic reflection. The reader may of course identify further themes. Those we have identified are grouped below into three categories, according to whether the emphasis is on mathematical ideas explicitly or implicitly in the curriculum, aspects of conceptual growth, or pedagogic approaches thought to be effective.

## Ideas throughout the curriculum

*Relations between quantities and properties* is a strong connecting theme in several chapters. Students are typically introduced initially to quantities. For example, geometry might focus on length, area, and volume in a direct way. The shift to relations between quantities may then seem like an obstacle. In geometry, for instance, the challenge in designing pedagogic approaches, including tasks, is how to direct student attention towards the properties of shapes to encourage the sort of deductive reasoning so valued in geometry, rather than solve problems by using imprecise measurement or trial-and-adjustment.

*Powerful implicit ideas* permeate the mathematics curriculum in such a way as to characterise what makes mathematics mathematical. Some of the most powerful ideas are so pervasive that they are not taught explicitly and yet are valued implicitly by the enculturated mathematician. We find these ideas appear in many of the chapters in this book, such as: variable, proportionality, similarity, symmetry, linearity, measure, dimensionality, representations, prediction, accuracy, discrete/continuous, transformation. The challenge for teaching is how to make these powerful ideas more explicit and joined-up, to show they are part of what it means to do mathematics. Indeed, these powerful ideas gradually become the focus in more advanced mathematics where the accent is placed on structure, for example linearity becomes a major area of study in linear algebra. These key ideas are important in employment too where students need to apply their mathematical thinking to novel situations. For example, recognising proportionality enables people to deal with scaling quantities.

*Formalisation* is a strong organizing theme throughout these chapters. This stage of schooling is characterised by a shift from formalisations of everyday understandings to learning new formal ideas that are hard to see in everyday situations. For younger students, everyday knowledge is a starting point for sense-making and then ideas are extended to less experiential situations. In a contrasting way, older students have to be introduced to formal ideas that are less intuitive and experiential but provide new ways to look at phenomena in and outside mathematics. Perhaps as a result, research about learning at this level is often about typical errors which are often reasonable consequences of particular designed learning environments. In this sense, there are no absolutes about learning at this level because much depends on curriculum aims, topic order, tool-use, and pedagogy.

*A sustainable curriculum* from the point of view of value to the learner seems to be one that is powerful for later employment, study, and citizenship. The research reported here indicates the desirability of a qualitatively different curriculum from those that are restricted to testable topics marshalled into a traditional order.

## Conceptual growth

*Sources of confusion* are often understandable as products of a mind which is under or over-generalising past experience, or over-dependent on under-informed intuition, or not fluent with ambiguous notations that require careful interpretation and repeated experience. Learners typically respond to visual appearance, drawing on perceived appearances and rushing to use recently-met or embedded procedures. At the same time they tend to ignore, or not attend to, structures, relations, and meanings that may not be immediately visible in the situation.

*Multiple purposeful experiences* within and outside of mathematics seem to facilitate more meaningful and powerful mathematical thinking for adolescent learners. Such experience may often be messy and require extended tasks over a period of time. Recently constructed concepts may not be used spontaneously but are still available as resources given sufficient nurturing. The corollary to this is that if teaching limits experience to situations that are obvious, or sim-plistic, or that yield to everyday and informal reasoning, or to images that only work in limited domains, it is ultimately insufficient even if it is necessary as a motivational starting point.

*Representations are critical features of mathematics and key tools for mathematical learning*, particularly the use of linked multiple representations. Often conceptual growth seems to lie in recognising the power of a representation or the structural connections between representations. By using representations, students can begin to recognise that the underlying mathematical concept extends into unexpected domains.

*Multiplicative reasoning* is central to mathematics in all domains but many other forms of reasoning are also important: deductive (geometry), structural (algebra), statistical, probabilistic, estimating, predicting, hypothesising, axiomatic, transformational. As with many of these themes, these forms of reasoning tend to be implicit in mathematics and hence left, if not carefully handled, as mysteries for students to unravel.

## Teaching approaches

*Concept definitions need to be introduced alongside non-examples and boundary examples* so that students can distinguish what is and what is not related to a new idea. Such an approach can counteract the apparently natural tendency to cling to prototypical examples when reasoning about space and functions. Prototypical examples are essential resources for everyone but learning involves recognising their scope and limitations. An example of the latter was when we observed a lesson in which a student posed the question 'is a quadratic in which the coefficient of the $x^2$ term is zero still a quadratic?' and time was given to discussing this.

*By exercising control over carefully designed software*, students can ask new powerful questions and engage more fully in enquiry-based approaches, developing skills for employment and future study. For example, when students use Logo, they might be allowed great control over the project direction. To develop their project, they would probably need to address the meaning of variable as a way of developing more general procedures that can be used efficiently. Using mathematics as a means of problem-solving becomes a highly important skill in employment and future study.

*Graphical representation* is a tool for learning in almost all areas of mathematics and teaching approaches need to facilitate connections between these and other representations such as algebraic, tabular, and computational. Students can develop the habit of graphing the data generated by situations and analysing the behaviour in terms of a repertoire of functions and other mathematical tools.

## A note on language

Before we go much further we need to clarify some meanings. There are many words used to describe the key beneficiaries of the education system: students, learners, children. There are also multiple ways of expressing the age or phase of schooling: elementary or primary for younger students; secondary or high school for older students. We are also aware that it can be irritating to describe 16–18-year-olds as *children* and just as annoying to describe 6-year-olds as *students*. We have chosen to use whenever possible the actual ages or age range so that an international audience is not distracted by local terminology. When it is too clumsy to do so we have usually referred to *children* below age 11 and *students* at age 11 and above and we also use 'learners' when speaking more generally. Our terminology for stages of schooling follows the UK pattern of 'primary' up to age 11 and 'secondary' from 11 to 18.

The above paragraph refers to 'we' as the authors. We have tried to maintain this use of the first person plural consistently to avoid ambiguity and all statements in this book have the agreement of all three of us. There are a few occasions on which we report on the particular work of only one of us.

## The web resource

Alongside this text, we have developed with the support of Nuffield Foundation and Oxford University Press a resource to support understanding of the issues raised in this book.

 www.nuffieldfoundation.org/key-ideas-teaching-mathematics

The web resource has a bi-directional relationship with this book. Thus, readers of this text may find it useful to seek out exemplification through the website, and we provide QR codes to aid this process. On the other hand, teachers may access the website first, either because they are directly looking for resources and ideas from this website or because they have been redirected from another website. In working with the resources that they find, some teachers may wish to explore further by reading the book to examine critically the theoretical underpinnings that have informed the design of those resources and ideas.

To create the website, we have trawled various sources and identified tasks which exemplify the key ideas presented in the text. In each case, the exemplification is supported by a brief commentary, though the website user would need to refer to the full text in this book for a critical discussion.

For example, in Chapter 2, it is suggested that algebra can be used to find out more about situations in and out of mathematics. Students can learn to use algebra successfully if they have multiple experiences, over time, of modelling situations in and out of mathematics, exploring and explaining what happens using various representations, and relating these back to the situation. There is a link to a task that demonstrates how various representations might indeed be used with this aim in mind. Research shows that having personal control of digital technology is important as it gives students a way to vary expressions, variables, and parameters and see the effects.

## An overview of the chapters in this book

There are differences in approach across the chapters. This is largely because existing research is not equally advanced in all domains. For example, algebraic thinking and proportional reasoning have been a focus of research activity over many decades so the relevant chapters are long with some tabular summaries. On the other hand research in statistics and probability education is relatively undeveloped, reflecting the fact that these areas of mathematics are more modern inventions and their place in the curricula of different countries varies widely. Furthermore, developments in technology are revolutionising the teaching and learning of statistics and so there is a need to focus on debates around these innovations. Chapters 6 and 7 reflect these recent developments.

Different chapters were led by authors with particular expertise in that knowledge domain so, although the authors have worked closely together, there are inevitable style differences, which we saw as *colour* rather than *inconsistency*. On the other hand we have tried to organise each chapter in a roughly consistent way. Each chapter addresses a content area of school mathematics through our take on what is important according to the research in that content area; this may not align in a transparent way with that in national curricula.

Although each chapter focuses on students' typical reasoning from age 11–16, consideration is nevertheless given to earlier mathematical experience and, especially in Chapter 9, to some more complex areas of mathematics, typically not addressed until higher education. This is why we have put '9–19' in the title, to indicate that the research has implications for teaching beyond the 11–16 range.

Each chapter contains discussion about models for progression though these are treated critically. It is inevitably difficult to disentangle what is the consequence of the curriculum and what might point to a need for more effective curricula and assessment in the future. Each chapter pays attention to what is reported in the literature about teaching approaches. In some chapters this is presented as a range of options with apparent consequences for choosing those options rather than as a recommended method. In other chapters, technological innovation is pointing to the need for rapid change in curricula and in these cases this research is reported in more detail without as much concern for balance.

In every topic area, more knowledge is needed about how students' understanding grows through teaching and learning in educational contexts. Each chapter ends with a section on where additional evidence is needed. Often, practitioners are in a position to contribute to the community effort in gathering such information. Therefore, where appropriate, we propose projects that could take the form of action research or collaboration in an active mathematics department.

## Chapter 2: relations between quantities and algebraic expressions

In the past school algebra has been concerned mainly with transforming expressions and equations so they could be more easily used, and relating these to graphs. Many of these tasks can now be done by software, but people still need to construct and recognise algebraic statements in their various forms. In addition there are new educational tasks: to use algebraic methods to model, understand, control, and make predictions about relations between quantities and variables. This chapter surveys the research on learning and evaluates different approaches to teaching algebra, from a focus on readily available software to a focus on fluent interpretation and use of the notational system. At the heart of all approaches is the need to recognise and move between equivalent expressions, equations and representations. There is no 'best way' to teach and learn algebra; choices about teaching need to be informed by the curriculum aims.

## Chapter 3: ratio and proportional reasoning

This chapter uses research about students' understanding of ratio and proportional reasoning to shed light on their prior understanding and also on the difficulties they may have in learning more. It outlines the relevant understandings students might bring with them from primary school and describes what is known about learning ratio and proportional reasoning for older school students. Proportionality and some related concepts appear everywhere in the curriculum but are often treated implicitly, such as when learning about measure, trigonometry, or gradients. Various teaching approaches are examined and suggestions made about how teachers might work with colleagues on teaching ratio and proportional reasoning coherently throughout mathematics.

## Chapter 4: connecting measurement and decimals

Measurement is a topic that connects and enriches the two crucial mathematical domains of geometry and number, and thence algebra through measurement formulae and relations between quantities. While measurement is sometimes allied closely with geometry in the curriculum, in this chapter we argue for a repositioning of measurement more with decimal representations than with quantification of spatial characteristics. The research reviewed in the chapter demonstrates that neither measurement nor decimals are best taught as simple skills; rather, the evidence points to ways in which each is a complex combination of concepts and skills that develop over a number of years.

## Chapter 5: spatial and geometrical reasoning

This chapter illustrates how geometry education, especially beyond the age of 9, needs to attend to two closely-entwined aspects of geometry across both 2D (plane) and 3D (solid) geometry: the spatial aspects, and the aspects that relate to reasoning with geometrical theory. The former involves spatial thinking and visualisation, while the latter involves deductive reasoning using approaches that employ, as appropriate, transformation and/or congruency arguments. The research reviewed in the chapter confirms that these twin aspects of geometry (the spatial and the deductive) are not separate; they are interlocked. It is through curriculum and teaching that bind together these twin aspects as a mutual whole that learners experience the full power of spatial and geometrical reasoning.

## Chapter 6: reasoning about data

Statistics in schools continues to focus around a given set of representations and procedures, which are often learned routinely and quickly forgotten.

That curriculum is however being challenged by statistical enquiry, which harnesses digital technology to explore data embracing a full investigative cycle. This chapter focuses on research into informal inferential reasoning, which will support teachers as the curriculum responds to the need for future citizens to reason effectively about data.

## Chapter 7: reasoning about uncertainty

One of the advantages of exploratory data analysis as described in Chapter 6 is that it avoids the known difficulties learners have with probability. However, as a result, probability is in danger of becoming isolated, trapped in a strange world of coins, dice, and spinners. Yet probability is a vital tool, not only for judging the reasonableness of pattern and trends apparent in data but also for judgements of risk by the ordinary citizen. This chapter reviews research about perceptions of randomness and probability, recognising the need for probability to be re-connected to the mathematics curriculum.

## Chapter 8: functional relations between variables

This chapter builds on Chapter 2 by relating equations, graphs, and functions, assuming some understanding of the notational system and its purposes. Ways to learn how to solve algebraic equations are evaluated, with descriptions of the possible strengths, limitations, and confusions associated with each method. The chapter summarises standard difficulties associated with graphs and graphing which may be problems of conceptual understanding rather than technical problems. A full understanding of functions goes beyond connecting equations and graphs, and can take many years and a variety of experiences to develop. Learners have to meet and use a wide range of functions: continuous and discrete; with and without time on the x-axis; smooth and not smooth; calculable and non-calculable. There are several distinct shifts of perspective that need to be made, and some research is summarised about associated problems, but there is no research that indicates a 'best' curriculum ordering.

## Chapter 9: moving to mathematics beyond age 16

As students make the transition to mathematics beyond the age of 16 the mathematics they learn brings together and extends the range of mathematical ideas they encountered earlier on in their mathematical career; in other words, the ideas covered in the earlier chapters of this book. This chapter gives a brief account of a careful selection of three of the new mathematical ideas developed at this stage: trigonometric functions, calculus/analysis, and statistical inference. We chose them as indicative examples of how earlier learning can be a crucial factor in laying secure foundations of understanding for further study.

The chapter concludes by returning to the theme of teaching for conceptual growth through a focus on powerful mathematical ideas.

CHAPTER 2

# Relations between quantities and algebraic expressions

## Introduction

Algebra is one of the most substantial areas of research in mathematics education. To keep the chapter to a manageable length there are some omissions, and some ideas have had to be compressed to be included and can be followed up in more detail using the references.

The literature is in strong agreement about the central purpose of algebra in the school curriculum. Algebraic manipulation without any meaning or purpose is a source of mystery, confusion, and disaffection for adolescents. 'Meaning' in school algebra comes from the way relations between quantities and variables are expressed. 'Relations between quantities' and 'algebraic reasoning' – used in the title of this chapter – pervade mathematics. Manipulating algebraic expressions enables us to express mathematical relations in different ways, and know more about them, but – possibly because fluent transformation is a valuable skill and easy to test – it is manipulation without meaning that tends to dominate the public view of the school curriculum. Teaching methods in school algebra need to bridge students' understandings of number to the use of algebra as a medium for mathematical reasoning.

We talk first about relations between quantities and algebraic reasoning, then about the prior understandings students might have, then about learning and routes of progression. Following that, the major part of the chapter summarises what is known from research about a number of different teaching approaches so that teachers can make informed choices. Because algebra is a much-researched field, this section is presented in tabular form. Finally, we suggest possible projects for teachers and key reading.

## The nature of algebra

Algebraic reasoning involves:

- formulating, transforming, and understanding generalisations of numerical and spatial situations and relations;
- using symbolic models to predict and explain mathematical and other situations;
- controlling, using, understanding, and adapting spreadsheet, graphing, programming and database software (based on Mason and Sutherland, 2002).

Employers and universities recognise these as goals for the curriculum. However, the detail of algebra as it appears in textbooks is often rather different, focusing mainly on formulating word problems and transforming by rearranging, factorising, and collecting like terms. Solving equations, which constitutes a large part of early algebra, straddles the first two of these statements while graphs and functions address all three.

Expressing school algebra in these three ways links it to what students know about relations between quantities. In their mental arithmetic methods, and in devising their own methods of calculating, students understand and use many of the fundamental relations that are later expressed algebraically. The simplest uses of algebra express relations between numbers when students already understand the relations. For example, finding $x$ when $2 + x = 5$ involves the additive relation: $2 + 3 = 5$ ; similarly, $a + b - a = b$ expresses the relation that students can spot when asked to calculate $37 + 49 - 37$. Another example of thinking algebraically is that if I have more cake than you have, then there is a method of equalising our portions so long as we know the relation of 'difference'. Choosing to halve the 'difference' is a use of algebraic reasoning about quantities.

There is a further reason for describing algebraic reasoning with these three aspects: it takes account of the wide availability of symbolic manipulators which can transform expressions, solve equations, and carry out other algebraic techniques. Such software is freely available and will be used by students on the Internet. Its

www.nuffieldfoundation.org/algebra-1

 www.nuffieldfoundation.org/algebra-3

 www.nuffieldfoundation.org/algebra-2

 www.nuffieldfoundation.org/algebra-4

use will help them answer questions which reflect all three statements above rather than traditional algebra questions focusing on manipulation. Conversely, to use such software requires an understanding of what it does, and why, which can only arise from engaging in algebraic reasoning. Kaput (1989), who worked for decades with young children on algebraic understanding, with and without the support of digital technology, described algebra as the sign system in which reasoning about relations is expressed, and came to the conclusion that, like any language, it is best learnt through communicating meaningful statements.

## What do younger students learn?

By the age of 11, students know a lot about relations between quantities from their everyday experience and from the primary mathematics curriculum. First, they have used several numerical relations. Most students should have a good understanding of the additive relation. If three numbers are related by $a + b = c$, then students also know that:

$$b + a = c \; ; \; c = a + b \; ; \; c = b + a \; ; \; c - b = a \; ; \; c - a = b \; ; \; a = c - b \; ; \; b = c - a$$

They may not be able to talk explicitly about this knowledge (and it is unlikely they would be able to symbolise it) but they would have had experience of using these variations. Understanding typically progresses from knowing that all these versions apply in particular cases, to knowing they are true for *all* cases, to possibly expressing this iconically (in diagrams or words) for all numbers. A few, depending on past experience, might be able to express this relation algebraically. They may also know that $a + (b + c) = (a + b) + c$ but may get confused if there are some negatives around.

Multiplicatively, most would know that $ab = c$ implies $ba = c$ for whole numbers, so many would also know that this implies: $c = ab$; $c = ba$. Again, we are not suggesting they can express this symbolically. This knowledge may depend on whether they know multiplication facts well enough to recognise multiples. Students tend to be less fluent with multiplicative relations than with additive, and may not have a full understanding of how division acts as the inverse.

An emphasis in primary mathematics on mental methods, derived number facts, and devising own methods may have led students to have some favourite relations they use when calculating. For example, they may know that adding the same number to two numbers does not affect their difference: $(a + d) - (b + d) = a - b$ and may be able to represent this on a number-line (Figure 2.1):

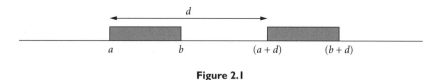

**Figure 2.1**

They may also know that multiplying a total by a scalar can be done after or before addition: $a(b + c) = ab + ac$. The use of 'chunking' larger numbers to aid calculation may have given them an intuitive understanding of distributivity and associativity. Some writers say that algebra is generalised arithmetic, in the sense that it is the language (or the sign system) in which we express such equivalences. Others say that the relations are fundamentally algebraic, and when we do arithmetic we are exemplifying algebra with particular numbers. Whatever your view, arithmetic, particularly mental arithmetic, provides a fertile ground for introducing algebra to express what learners already know about number relations.

Relations between quantities can be understood without knowing actual quantities. Schmittau (2005) shows that even very young children can reason about relations between unmeasured quantities if they can handle and compare them, such as when pouring and mixing drinks. This reasoning is also a basis for algebraic reasoning. For example, they know that if $a > b$ and $b > c$, then $a > c$ and they can express those relations even if they do not know the quantities $a$, $b$, and $c$. Similarly, they know that if $a = 2b$ and $b = 2c$ then $a = 4c$, and can also express relations such as when 5 cups of water are equal in quantity to 3 other-size cups of water ($3x = 5y$) (see also Davydov, 1990; Dougherty, 2008; Kaput, 1998, 1999). Schmittau and other teachers following similar methods have found that young children can learn to use algebraic symbols to express these relations if they are introduced as a way to record relations between quantities, and this use is enculturated, over time, into the regular work of mathematics lessons.

On entry to new schools in early adolescence, therefore, students may have significant knowledge about relations between quantities. Some students may have expressed these algebraically, or may have met algebraic formulae such as $l \times b$ for area of a rectangle of length $l$ and breadth $b$. The use of conversions and formulae for area, perimeter, and volume, are useful pre-cursors for algebra (NMAP, 2008). In science, and maybe in mathematics, students may have had experience of expressing relations between variables graphically, such as temperature against time, or distance against time, or height of a growing plant against amount of plant-food, or costs of school events against number of students participating, although apart from time the quantity in the x-axis may not have been continuous.

Students may also be able to handle 'hidden number' problems in which unknown numbers complete a number sentence but these may have been limited to cases similar to $n + 2 = 5$, which can be solved using remembered number facts, rather than those that need algebraic methods such as $3n + 11 = 5$.

All these experiences relate to the three aspects of algebra with which we started. Algebra teaching has the task of joining these earlier school experiences to the goal of algebraic reasoning, with the added task of developing the technical fluency that makes this linkage possible. Traditionally, the focus has been on the latter rather than the former.

## Students' algebra learning

The way young children learn about numbers and quantities is well-documented (e.g. Nunes *et al.*, 2009) but most research on algebra focuses on students' errors, that is what they cannot do, rather than what they can, or on what students can do in particular circumstances. Nevertheless, looking at typical errors provides a window on how students can experience algebra. For example, many errors arise from weak understanding of the notation (e.g. MacGregor and Stacey, 1997; Stacey, 1989) and another cause of error is the misuse or misremembering of techniques. The history of algebra teaching and learning shows us that emphasis on abstract manipulations, divorced from relational meaning, can lead to error-prone work and also to disaffection with mathematics (e.g. Kieran, 1992). A further source of error is due to ambiguities within algebra, and particularly possible gaps between arithmetical and algebraic reasoning and notation.

Finally, students often hastily apply the last thing they learnt, or respond to visual appearance, rather than 'read' the meaning of the algebraic statement in terms of the relations it represents.

## Problems with the interpretation of notation

From the CSMS (Concepts in Secondary Mathematics and Science) study of 3000 students in the 1970s, Küchemann (1981) reports a range of ways they treat letters in questions about understanding and using algebraic notation, collecting simple like terms, and expressing relations. They may

- evaluate letters in some way, for example $a = 1$;
- ignore letters, for example $3a$ taken to be 3;
- treat letters as shorthand, for example $a =$ apple;
- treat letters merely as objects to be moved around;
- use letters as specific unknowns to be found;
- use letters as generalised numbers;
- use letters as variables.

In a follow-on study Booth (1984) had further insights into sources of confusion arising from notation:

- some students believe that different letters have to have different values, so would not accept $x = y = 1$ as a solution to $3x + 5y = 8$;
- students might believe that a letter cannot have different values in the same problem, such as when an equation has multiple roots;
- students might expect the same letter to have the same value in different problems;
- the use of coding to introduce algebra can lead students to expect values that are related to the alphabet ($a = 1, b = 2$ ...; or $y > p$ because of relative alphabetic position);
- letters in algebra do not stand for objects (e.g. $a$ for apples) but are used for units (such as $m$ for metres);
- some textbooks always arrange letters in algebraic order, but this is not an essential aspect of algebra (e.g. $px + qy = k$);
- different symbolic rules apply in algebra and arithmetic, for example '2 lots of $x$' is written '$2x$' but two lots of 7 are not written '27'.

Booth's and Küchemann's work shows that algebraic conventions are not obvious. There are several inherent problems that are often glossed over in textbooks and curricula. As well as needing to understand '=' as 'is equal to' rather than 'makes', students have to know further relational meanings: equivalence and identity. They have to know that $3(x + 2) = 3x + 6$ and $3(x + 2) = 2x - 7$ involve two different meanings of '=', the first being 'is equivalent to or identical with' and the second being 'is equal to for some value of $x$'. A further problem is that letters can stand for labels, givens, unknowns, variables, parameters, or constants, and some letters have specific meanings. Students need to see a difference between formulae (connecting quantities), equations ($x$ as unknown), identity ($x$ as the argument of function), properties (generalizing patterns), and functions (relations between variables) (Usiskin, 1988). They also need to experience a wide range of algebraic actions: translating, transforming, generalising, solving, simplifying, graphing, justifying, and expressing relations and structures (Usiskin, 1988). Merely knowing that 'letters stand for numbers' is not enough.

www.nuffieldfoundation.org/algebra-3

www.nuffieldfoundation.org/ algebra-6

www.nuffieldfoundation.org/algebra-5

www.nuffieldfoundation.org/ algebra-7

Some authors (Filloy and Rojano, 1989; Linchevski and Hersovics, 1996) talk about a significant gap between arithmetic and algebra arising from some of the issues raised above. To practise algebraic methods, such as inverses, transformation, and solution algorithms, students need to work on examples which cannot be solved using arithmetical facts alone. Students have problems solving linear equations where the solution is negative or a fraction because they often try to use number facts, or trial-and-adjustment, rather than using algebraic methods, such as inverses. If they do not shift from doing arithmetic to doing algebraic transformation they run into difficulties later. Without grasping the need for algebraic transformation, the methods needed for further work, such as simultaneous equations or quadratic functions, can appear to be a bunch of unrelated techniques to remember. However, the idea of a 'gap' is one of definition. If algebra is taught mainly as a set of methods of transformation then there will be gaps where methods are not obviously connected; if algebra is taught through the construction of meaningful expressions and equations, then knowing when there is only one value for $x$ and how to find it can be seen as algebraic reasoning. Boero (2001) considers that learning transformation processes is only purposeful as part of problem-solving, modelling, proving, and conjecturing activity (p. 99). As such it requires students to anticipate what can be achieved through transformation. He claims that learning routines can even hinder meaningful learning because it encourages actions without anticipating their consequences.

To illustrate the complexity of notational understanding, Sfard and Linchevski (1994) pose the question:

For what values of the parameters $p$ and $q$ does the equation $(p+2q)x^2+x=5x^2+(3p-q)x$ hold true for every value of $x$?

Solving this requires very careful distinctions to be drawn between the meanings of letters. Knowing how to solve equations is of little help on its own; first of all suitable equations have to be set up from the information in the question. The problem has to be approached from the point of view of understanding its structure, not as a string of operations, nor as a quadratic in $x$ because we are told this is true for *all* values. We have to make the coefficients equal for both sides to be identical. Sfard and Linchevski argue that the shift of perspective students have to make in order to work with functions like this is smoother if students first understand the difference between parameters and variables, otherwise there will be clashes with their arithmetical (operational) understanding. They say: 'symbols do not speak for themselves' and that students need to know what to look for, identifying two shifts of difficulty. The first shift is from reading algebraic expressions as a string of operations to understanding that they show how the variables are related and how the parameters form the shape of this relation. The second shift is from seeing letters as unknown numbers to seeing them as variables. Arcavi (1994) describes this as the development of 'symbol sense' and an ability to 'read through symbols'.

As we shall see, however, this insight about reading symbols does not imply a 'best' teaching sequence.

## Research into teaching and learning approaches in algebra

We shall now look at several approaches to researching 'what works' in learning algebra, but 'what works' always depends on the goals. It has always been the case that *some* students can and do learn algebra successfully as a toolbox of techniques for later use, although it is typical to make errors such as being inaccurate with minus signs. Research about repetition and memory shows that reliance on these alone leaves most learners likely to make mistakes and to be unable to adapt. Algebra taught as a tool to express and transform situations leads students to use transformations meaningfully because they have to think about what sort of expression they need in order to solve a problem or prove a result.

## Learning manipulation techniques

A major problem in early algebra is students' desire to conjoin terms in an effort to make an 'answer', for example $3a + 2b = 5ab$. Tirosh *et al.* (1998) watched two experienced teachers trying to help their students avoid conjoining. One teacher took a ritual approach, referring again and again to 'like' and 'unlike' terms, using lots of examples for practice and generalisation of procedures. Her students did, in the short term, adopt correct rules for simplifying algebra. The other used a conflict approach, using substitution to compare the effects of doing operations in different orders and combinations. Her students learnt about non-equivalence and also gained some sense of an expression as an object in its own right. It depends whether you value fluency over understanding, or vice versa, which lesson can be seen as more successful. Thomas and Tall (2001) report that their students did not conjoin after they used a computer algebra system (CAS) to manipulate and evaluate expressions. The CAS gave insight into the conventions of procedural and substitution methods, but lacked the conflict of Tirosh's teacher which emphasised the *reason* for manipulation rules, and also lacked the fluent practice provided by Tirosh's more dictatorial teacher.

Another approach is to work on the meaning of '=', so that arithmetic becomes the study of relations among numbers rather than purely about computation.

 www.nuffieldfoundation.org/algebra-1

 www.nuffieldfoundation.org/algebra-3

www.nuffieldfoundation.org/algebra-2

Jones designed software which enabled pre-algebra and early algebra students to substitute number expressions into equations, such as being given 49 + 6 = 52 + 3 and several other statements including 52 + 3 = 55. By physically moving the latter into the former on the screen they constructed the new statement 49 + 6 = 55. Through this kind of task they learnt to treat equations as objects, expressions as terms, and substitution as a method of transformation (Jones, 2008; Jones and Pratt, 2006, 2012). To see algebraic symbolisation as a purposeful problem-solving tool, Sutherland and Rojano developed a spreadsheet-algebraic approach (1993). Students solved problems by using the cells of a spreadsheet to hold the unknowns and then they built up the relationships expressed in the problem within other cells. In this approach students progressively modified the values of the unknowns until some target totals were reached. In Mexico and England 15-year-old students successfully learnt to use the spreadsheet to set up relevant equations using *ad hoc* exploration. Wilson *et al.* (2005) developed this idea further, designing tasks that promoted a need for algebraic symbolisation so that students could understand the lasting utility of algebraic ideas, rather than merely use them for ephemeral purposes. For example, they used the Fairground Game in which students had to arrange numbers 1 to 5 in the left spreadsheet column so that they could achieve the total in the bottom spreadsheet cell (see Figure 2.2).

| ◇ | A | B | C | D | E |
|---|---|---|---|---|---|
| 1 | 1 | | | | |
| 2 | 3 | 4 | | | |
| 3 | 5 | 8 | 12 | | |
| 4 | 4 | 9 | 17 | 29 | |
| 5 | 2 | 6 | 15 | 32 | 61 |

**Figure 2.2** Getting the total 61 in the Fairground Game (Excel screen).

This final understanding took time to achieve, and some students focused only on using algebra to get a particular answer rather than on the power of the method for getting answers more generally. In an answer-orientated classroom this difference would affect students' willingness to engage with algebra. Using algebra to control software appeals to the adolescent need for autonomy and power, where following rules without purpose does not.

Computer algebra systems (CAS, such as Mathematica™ or Maple™) that can manipulate expressions alter students' access to algebraic reasoning. Thomas and Tall (2001) suggest that students can combine fluent transformational performance with CAS, which demonstrates the conventions, if appropriate tasks are given which require both. As with all tools, the pedagogy matters – the tools have to be understood as tools *for* something. Hembree and Dessert (1986) report that students who regularly use hand-held technology focus more on the underlying concepts and can be just as fluent with techniques as non-users, possibly because of their increased familiarity with what algebraic manipulation ought to look like. Horton *et al.* (2004) got some students to follow step by step tutorials in a CAS system for solving linear equations and compared their work to a control group who used calculators to calculate, but not to solve. Medium to high achieving students did better after CAS use than the control group, and they were better at solving problems with fractions and several steps, maybe because they had experienced solving as a process rather than as an arithmetical search. However, lower achieving students did not do better after CAS-tutoring as they tended to use it as a crutch rather than as source of information and feedback. CAS use had enabled most, but not all, students to understand the role and nature of coefficients and effects of changing them, the ultimate goal being not to use the tool, however, but to do the task fluently and accurately themselves.

A problem for those who believe that fluent practice leads to understanding is the need to explain how this happens. Davis (1985) expresses this as 'if the students spend enough time practicing dull, meaningless, incomprehensible little rituals... something WONDERFUL will happen' (p. 118). 'Something wonderful' *can* happen if routines become fully internalised and hence available for future use, such as happens with writing, reading, learning dance steps, and so on (Hewitt, 1996). All of these become worth learning if they are incorporated immediately into longer and more meaningful sequences of work. 'Something wonderful' can also happen when manipulations have additional physical components through speech patterns (Hewitt, 1996) or diagrams and on-screen actions (Jones, 2008) that provide sensory experiences that aid memory and add depth to repetitive fluency. 'Something wonderful' can happen further if there are relations to be seen in the outcomes of the work done. Many modern textbooks rarely vary questions in ways which draw attention to the relations and patterns being represented, such as relations between parameters and roots in quadratics, but older textbooks often did this so that there were properties for students to notice while working on exercises, for example by juxtaposing $x^2 + 5x + 6$; $x^2 - 5x + 6$; $x^2 - 5x - 6$; $x^2 + 5x - 6$. As with CAS use, students who focus on the detail of individual questions rather than the process and patterns may not gain understanding from practice alone.

## Learning higher school algebra

There is only a little research about how students learn the technical aspects of algebra beyond the elementary stages reported by Küchemann (1981), Booth (1984), and Hodgen *et al.* (2009). What there is supports the view that understanding and anticipating the purpose and relational meaning of manipulations makes a difference to learning for most students.

A well-known way to harness students' natural pattern-seeking is through seeing algebra as the construction of statements of generality. This has often been studied using pattern sequences, that is to express general functional rules that relate the *n*th term to its value. For example, the tabular form of data for a sequence of nested squares suggests that the sequence increases by 4 except in the first case (Figure 2.3).

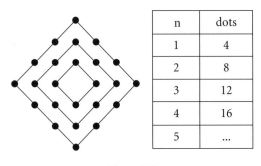

| n | dots |
|---|------|
| 1 | 4 |
| 2 | 8 |
| 3 | 12 |
| 4 | 16 |
| 5 | ... |

**Figure 2.3**

The number of dots for the nth square is therefore 4*n*.

Bishop (1997) describes the stages of understanding that middle school students have to go through in such situations: counting and verbal directions; relationships as single operations; term to term sequential relationships; functional relationships between sequence position and the value of the term. Lee (1996) reports that students who learnt some algebra this way, after earlier failures with a more procedural approach, were able to become fluent with procedures very quickly afterwards, motivated by understanding of their purpose. Radford (2008) also reports success, but research results with this method vary, and as with most methods the difference is probably due to the accompanying teaching and students' understanding of the aims and purpose of the tasks. The final shift to the functional relationship has to be supported somehow by the teaching, the task, or the technology as it requires a change of focus away from a tabular approach.

Students often use visual clues to trigger their methods and many classic errors are due to over-reliance on what things look like, for example interpreting $r^2$ as 'doubling $r$'. In the squares example above students may want to label the first square as '2' because of the number of dots on its sides. The most useful analogies are relational similarities, not perceptual similarities (English and Sharry, 1996). Searching for similar relations and structures among visually dissimilar expressions is a key aspect of becoming proficient in algebra. Vosniadou and Ortony (1989) describe tasks which emphasise 'salient similarity' to be able to make these kinds of distinctions. For example, students eventually might be able to identify inverse proportionality in an equation such as $m = 1/r^2$ by seeing $r^2$ as a variable, or linearity in $\cos 2\theta = 2 \cos^2 \theta - 1$ by seeing $\cos 2\theta$ and $\cos^2 \theta$ as variables, if they have learned to focus on structure rather than actions. Seeing similarity where teachers hope it will be seen is not immediate for most students; they need to experience a neighbourhood of similar structures over time in order to learn how to search beyond obvious visible signs.

# Progression

## Curriculum content and organisation

As workplace and academic needs for algebraic reasoning expand, curriculum developers have tried different approaches to make algebraic ideas more accessible in school. Expression of generality, modelling realistic situations, problem-solving, constructing algorithms, and graphical functions have all been developed as approaches to develop algebraic reasoning. The extensive work of Yerushalmy (2001, 2005) and Kaput (e.g. 1998) over time shows that students who first encounter algebra as a way of expressing patterns, relations, and functions that they already understand and then manipulate, through software, can become fluent and use it with power and relational meaning. Yerushalmy and Kaput totally reorganised the usual curriculum progression from arithmetic to algebraic manipulation, then to graphs and functions, to show that progression is dependent on curriculum and teaching. In their programmes, arithmetic, algebra, data, graphs, and functions co-exist and represent the same situations side by side, and students can move between them as they choose[1] (Yerushalmy et al., 2002).

---

[1] These projects also show what can be achieved when a curriculum is not rigidly prescribed.

By contrast, some other researchers and projects aim to improve acquisition of manipulative and transformative skills, such as with fine-tuned responsive learning programmes (e.g. Anderson and Fincham, 1994), concrete and visual models (e.g. use of the balance metaphor), and so on.

How can teachers choose between these multiple pathways? One way to think about school algebra would be to consider the opportunities offered by each approach, and we do this below. Another way would be to ask – what aspects of algebraic reasoning *should* students experience? Bednarz *et al.* (1996) argue for a balanced coordination of various experiences (see also Bednarz and Janvier, 1996; Bell, 1996). Kaput (1989) explains how this can be done with curriculum sequencing that combines procedures and concepts through consolidation, simplification, and generalisation. The generalisations then become new mathematical objects which act in more abstract systems. For example, in the classic handshakes problem (if *n* people all shake hands with each other, how many handshakes will there be?) children try various numbers of people shaking hands, find a way to present this structure, then generalise according to the patterns they find. Older children might be encouraged to predict numbers by looking at the structures of the handshake patterns. This produces a formula (maybe in the form of spreadsheet instructions) which will work for any number of people – a new mathematical object. This formula can then become a focus for enquiry: How quickly do the numbers grow? Do we get zero for one person? Can we find the number of people who will take the number of handshakes over 100? and so on. Development through this kind of activity depends on repeated experiences, so that algebraic reasoning becomes one of several tools that teachers and students use in any mathematical context. For example, Kaput's team uses a cyclic development of algebra as involving:

- transformations: for example changing the subject, equivalent expressions;
- comparing representations: for example comparing a formula with a graph;
- producing parallel descriptions using non-symbolic systems such as everyday language and pictorial images (in the handshake example this might include creating diagrams to explain and justify different versions of the formula).

They had significant success in teaching young children to use algebra with multiple representations in which there were repeated experiences involving describing situations, retrieving data, graphing, suggesting formulae, fitting and adjusting curves, and so on. Rules and manipulation came about as generalisations of actions in situations.

Many textbooks offer two progression routes for algebra. One moves from linear sequences to representing sequences on coordinate axes, to describing them as linear functions, and hence to quadratics and cubics – the underlying idea being the expression of generalities about number patterns. The other moves from arithmetical laws to the syntax of algebraic expressions, with transformation being introduced to enable equations to be solved, finally joining the other route with graphical methods of solving higher order equations. Even where these routes are conceptually coherent in textbooks, it is still possible to emphasise technical aspects over purpose so that learners do not understand the development. Kaput's approach through extended task sequences aims to join these two possible progressions.

## Mathematical aims for an algebra curriculum beyond modelling

The developmental paths described so far all suggest modelling (expressing situations as manipulable functions) to be the ultimate aim of school algebra, and this accords with the needs of employers and many higher education routes. However, the ability to express unfamiliar situations algebraically as relations, and to transform those relations in order to act further on them in efficient ways, is also at the heart of advanced pure mathematics. The construction of proof, and the use of abstract algebra to express objects which cannot be represented by single numbers (e.g. vectors, matrices, groups), depend on competent understanding and use of symbolism. Jones' work with middle-school-aged children shows that they are willing and able to work within mathematical systems that have no apparent external usefulness but are internally axiomatic and coherent (2008).

Fey (1990) and Arcavi (1994) have described aims for algebra curricula which are compatible with 'modelling' aims, but which also address the needs of pure mathematics. Fey focuses on the necessary abilities to:

- estimate patterns that could emerge from a given algebraic expression in numbers or graphs;

- compare orders of functions;
- conjecture likely algebraic expressions that would describe a given set of values or a given graph;
- choose the most appropriate of several equivalent forms of a relation for a given purpose;
- anticipate the result of an algebraic operation.

In contrast Arcavi talks of the impossibility of ever defining a complete list of what it means to have symbol sense, which for him is a collection of dispositions deriving from a range of experiences in school algebra. The dispositions include: being on friendly terms with symbols; being able to read them and manipulate them in order to read them better; being able to engineer them for a purpose; and knowing when they are equivalent. To get to this state, students need a range of complex learning experiences which coordinate the use of symbols, manipulating them, making sense of them, using tools to free up the mind to make meanings and connections, having generalities to express, evaluating methods and solutions, and extending their experience with 'what if' questions.

## Teaching approaches

Having established that there are choices to be made about curriculum aims, content, and sequencing, we now analyse several possible teaching approaches:

- create a need for algebra to express equivalent arithmetical relations;
- express relations among quantities;
- create a need for algebra to model phenomena;
- learn about algebraic expressions by substituting numbers in them;
- learn about expressions by 'reading' them;
- learn about expressions by modelling them diagrammatically;
- transform expressions by collecting like terms, factorising, simplifying;
- construction tasks.

Teachers can use these analyses to reflect on strengths and limitations, and to identify approaches which might supplement, or lead them to alter, their current methods possibly using the special cases and links we suggest. In each case we evaluate their possibilities and limitations. The limitations of each method might prompt teachers to devise tasks to find out if their own students exhibit these problems.

# Create a need for algebra to express equivalent arithmetical relations

Equivalent expressions $4a + 4 \equiv 4(a + 1) \equiv 2(a + 2) + 2a \equiv 4(a + 2) - 4$ can arise from counting paving slabs round an $a$ by $a$ square in different ways.

Equivalent equations $x = 3; x + 2 = 5; 2x - 3 = 3; 7 = 2x + 1$ can arise from classroom games to 'hide the number'.

## ASSUMPTIONS ABOUT PRIOR EXPERIENCE

Students understand that arithmetical statements express relations, and are not always instructions to calculate: e.g. they know that $7 - 5 = 2$ is another way to write $5 + 2 = 7$. They know about 'doing and undoing' in arithmetic, and the use of brackets to contain numerical expressions. They have some mental arithmetic.

## POSSIBILITIES OF THIS APPROACH

If students understand the underlying relations, symbolic notation is meaningful. For example, $3a + 3b$ means three of quantity $a$ added to three of quantity $b$. They might express a situation as $3a + 3b$ or $3(a + b)$ and see that these are equivalent. They might learn to think about relational meaning before acting on expressions.

Fractions, decimals, and variables can be used in such expressions, and fractions automatically arise in expressing multiplicative relations.

Mental methods can be expressed as general rules. Understanding equivalence arises from class discussion about different ways to express situations.

This approach provides a foundation for understanding variables and functions and for programming algorithmic software. Multivariate situations are accessible such as: $p + q = 10$ can describe two numbers that total 10, so if I know $p$ I can use $10 - p$ to find $q$.

## LIMITATIONS OF APPROACH

Algebra as 'generalised arithmetic' can lead to common misunderstandings about notation. Thinking about relational meaning can prevent some of these, but can slow fluency.

## SPECIAL CASES AND LINKS TO CONSIDER WHEN USING THIS APPROACH

Comparing equivalent expressions helps to understand how to transform them.

Explicit work on the meaning of '=' is necessary.

Addition and subtraction are inverses in the additive relation; multiplication and division are inverses in the multiplicative relation.

There needs to be attention given to less obvious notations, such as $2r$ and $r^2$.

Students need to consider the domain of application of the relation.

Use ICT to explore equivalent expressions and discuss the most efficient.

Order of operations is decided by the meaning of the notation. Rules can be misapplied (e.g. in $8 - 5 + 2$ to give 1 instead of 5).

# Express relations among quantities

### EXAMPLE

$p + q = 10$ can describe the capacities of pairs of containers whose contents total 10 units, so if I know $p$ I can use $10 - p$ to find $q$.

### ASSUMPTIONS ABOUT PRIOR EXPERIENCE

Students understand the nature and dimensions of the quantities, for example length, capacity, and other physical properties. Students can express $>$, $<$, and $=$ relations. They can describe addition of quantities and saying how much or how many of one measure go into another. They may know some equivalent measures.

### POSSIBILITIES OF THIS APPROACH

Algebra expresses situations they already understand. It can express relations between quantities when we do not know the individual quantities, such as using $t + 10$ to describe 10 minutes after time $t$, or $b - 5$ to describe the age of someone 5 years younger than someone of age $b$. In these cases, the letter could be a parameter of the relationship or an unknown to be found.

In converting measures, algebra is useful because it expresses relations between variables. A need to use fractions and decimals arises from multiplicative relations and from expressing one quantity in terms of another.

This approach provides a foundation for understanding variables and functions, and also for modelling realistic situations and for programming algorithmic software.

### LIMITATIONS OF APPROACH

Expressions of order one, two, and three, or exponential expressions, can be generated using such approaches, but higher order polynomials are harder to see as combinations of quantities. Area models for quadratics and scalar multiples are effective, volume and capacity for cubics. Negative numbers are hard to 'see'. Some quantities: area, time, angles, temperature, and negative measures, require imagination.

Letter use can confuse, for example is mg '$m$ grammes' or 'milligrammes' or 'mass times acceleration due to gravity'? In the formula $A = \frac{1}{2}bh$ are all three letters variables? Does $A$ stand for a given or a variable area? Or is it a way of 'seeing' the area of any triangle? Or is it an algorithm for calculating area?

### SPECIAL CASES AND LINKS TO CONSIDER WHEN USING THIS APPROACH

Students need to: develop fluency of manipulation and use of notation conventions; see similar relations arising in physically different situations; connect symbols to meaningful graphical representations; appreciate dimensionality; use ICT to explore equivalent expressions; devise the most efficient expressions using fewest variables.

# Create a need for algebra to model phenomena

## EXAMPLE

How much does a beanstalk grow in a day? How high am I after turning through $x$ degrees on a fairground wheel?

## ASSUMPTIONS ABOUT PRIOR EXPERIENCE

Students have experience of different kinds of covariation phenomena and a repertoire of ways to represent them: flow diagrams, equations, formulae, graphs etc. They understand the meaning of algebraic expressions met so far.

## POSSIBILITIES OF THIS APPROACH

There is a need to identify variables and minimise the number of variables used. Algebra is needed to describe and predict, so students appreciate its power. Real phenomena can be used. Algebra has a purpose and students can relate empirical experience to formal mathematics, maybe controlling a spreadsheet to store and predict events. Covariation can be expressed algebraically and graphically. The formal relations that underlie existing models, for example physical 'laws'; traffic light systems; savings accounts, can be understood. Dimensionality can be addressed and there is the potential to discuss rates of change, turning points, maxima and minima.

## LIMITATIONS OF APPROACH

There is a tendency to think additively, followed by tendency to assume linearity as students get older. There is a need to understand either the algebra or the situation well in order to connect the two and evaluate predictions.

Students tend to make *ad hoc* decisions that do not extend mathematical knowledge. There is no need to manipulate or simplify expressions or transform equations as the modelling software will do that.

Students tend to join up data points without considering whether the same relation applies across the whole domain, and without considering what happens at zero on the horizontal axis. Those who reason about possible relations do better than those who use a trial and adjustment approach to build algorithms.

## SPECIAL CASES AND LINKS TO CONSIDER WHEN USING THIS APPROACH

Students need several experiences with mathematically similar phenomena in order to get a usable understanding of abstract formal representations, to recognise similar structures in future, and to identify an efficient set of variables. These need to be carefully planned and sequenced and not left at an ad hoc 'problem-solving' state.

Multiple experiences of this kind of activity are necessary for general development of knowledge of graphs, functions and their expression in algebra or through statistical descriptions.

Understanding differences between empirical data and function-generated data.

# Learn about algebraic expressions by substituting numbers in them

What is 60 degrees Celsius in Fahrenheit? What about −40 degrees? When the temperature goes up one degree in Celsius, how far does it rise in Fahrenheit?

## ASSUMPTIONS ABOUT PRIOR EXPERIENCE

Students need to be fluent with elementary arithmetic such as number bonds and multiplication facts, unless they are using a pre-programmed algorithm.

## POSSIBILITIES OF THIS APPROACH

Comparing processes and outcomes of substitution in algebraic expressions is a good way to explore equivalence, equality, inequality, and differences between notations. Substitution focuses on individual points on graphs. Substitution, especially using ICT, shows whether an expression, formula, or a solution to an equation is likely to be correct. Input/output data pairs can focus students on 'what expression would have generated these pairs?' Counter-examples can be generated. Substitution of expressions, rather than numbers, exposes the structure of relations.

## LIMITATIONS OF APPROACH

Substitution of numbers as an exercise in itself emphasises the arithmetical aspects of algebra without focusing on the overall relations and structures. The main problems are with fractions and negative numbers, especially combinations like subtracting negative numbers. Problems with negative numbers may be due to misunderstandings, for example '− 5 − 3 = 8 because two negatives make a positive', or inaccuracies due to haste. The latter is common at all levels of mathematical work. Substitution exercises appear in textbooks without any clear purpose.

## SPECIAL CASES AND LINKS TO CONSIDER WHEN USING THIS APPROACH

If expressions have been generated from relations that students know about, misuse of notation is less likely to occur.

The outcomes of substitutions give information about the nature of the expression or equation.

Knowing that a graph consists of points that arise from a function or formula is not the same as knowing that ALL points in the graph satisfy the function, and ALL points that satisfy the function are in the graph.

Substitution of one expression for another, to simplify working, has more complex uses in integration, combining functions, changing bases, and so on.

# Learn about expressions by 'reading' them

### EXAMPLES

I think of a number, divide it by 5, subtract it from 2, square what I get, add 1 to it and square root everything and the outcome is equal to 4

Whatever the values of $x$ and $y$, $z$ will be three lots of $x$ added to two lots of $y$.

It is always true that the square of the total of two numbers is equal to the squares of the two numbers added together, plus twice their product.

### ASSUMPTIONS ABOUT PRIOR EXPERIENCE

Knowledge of arithmetical operations and the meaning of brackets and '='.

### POSSIBILITIES OF THIS APPROACH

Reading algebraic expressions as sequences of operations rehearses use of notation conventions and connects notation to mathematical meaning and the relations being expressed. It is less likely that students will do meaningless actions on expressions they understand.

### LIMITATIONS OF APPROACH

Gets rather complex if it is extended to think about higher order functions and equations, and multiple variables.

### SPECIAL CASES AND LINKS TO CONSIDER WHEN USING THIS APPROACH

It is worth returning to 'what does this bit of algebra mean?' when students have misused an action or algorithm.

Students need to think about relational meaning before applying any algebraic technique.

# Learn about expressions by modelling them diagrammatically

Rod diagrams and area diagrams (Figure 2.4):

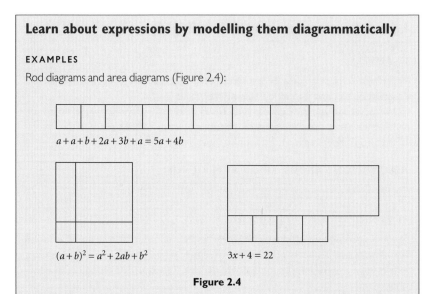

$a + a + b + 2a + 3b + a = 5a + 4b$

$(a + b)^2 = a^2 + 2ab + b^2$

$3x + 4 = 22$

**Figure 2.4**

### ASSUMPTIONS ABOUT PRIOR EXPERIENCE

Familiarity with images of objects being used and how they combine and extend.

### POSSIBILITIES OF THIS APPROACH

Cuisenaire rods, number-line, and area models represent quantities and are hence extendable for algebraic relations. They also provide physical experience of rearrangement and simplification, and an image to return to mentally in the future. Isomorphic models enable two-way reasoning, either with the model or with the symbols. For example, a balance model emphasises equality for linear equations in one unknown.

### LIMITATIONS OF APPROACH

Mental images are particularly helpful if they incorporate similar meanings to the mathematics they are representing, otherwise their use is limited and students may apply them inappropriately. Concrete and metaphoric models do not adapt well to negative numbers and sometimes do not adapt to fractions.

Object models do not represent quantities and may encourage the erroneous view that the letters are shorthand or code for words. They only represent adding and subtracting and scalar multiplication. If they are used to distinguish like terms students may think that 'likeness' depends on individual letters, rather than the form of terms. In algebra, $a$ and $b$ do not always represent different values.

### SPECIAL CASES AND LINKS TO CONSIDER WHEN USING THIS APPROACH

Major problems in algebra include: understanding when a letter represents an unknown, a variable, a particular constant (e.g. $\pi$), a parameter, or some other object such as a function. Algebra is not just arithmetic with letters, but is about relations among quantities and abstract numbers.

If visual images are used, students have to detach from these images which may become obsolete or limiting in more advanced contexts.

# Transform expressions by collecting like terms, factorising, simplifying

## EXAMPLES

Factorise, $ab + a^2b + ab^2 + b^2$

## ASSUMPTIONS ABOUT PRIOR EXPERIENCE

Students understand that letters stand for numbers, that expressions stand for relations between numbers, and that equations show that two expressions are always, or sometimes, equal in value. They understand what a 'term' is and that terms are distinguished not only by their letters but also by their form. It helps to know what an expression means.

## POSSIBILITIES OF THIS APPROACH

Programmed learning packages, or 'teach the computer' tasks, can be effective learning tools. Colouring, underlining, or circling particular terms can help distinguish 'like' from 'unlike' terms. Fluency with procedures can develop if exercises are well-designed and students can deal with negative signs. Students can learn about uses of algorithms and come to recognise equivalent expressions. They can learn how to find unknowns in fully-defined equations; to find where some functions cut the x-axis; and to transform expressions into more useful forms. They can understand the outputs of computer algebra systems and use transformations and substitutions to minimise number of variables necessary in modeling. This helps towards designing efficient algorithms and other methods.

## LIMITATIONS OF APPROACH

Students report that this aspect of algebra turns them off mathematics. Students who do not know why they are doing this, and how and when to check accuracy, may come to depend on memorised limited rules which they tend to misapply or misremember, for example 'change the side change the sign' does not work for inequalities, nor for multipliers or divisors; 'multiply the First, Outer, Inner, Last (FOIL)' does not work for more terms in brackets, nor for more bracketed terms. BODMAS and its variants do not work for some structures. Other temporary tricks, such as using $a$ for apple and $b$ for banana when collecting like terms, are also limited in application and detract from the meaning of number.

## SPECIAL CASES AND LINKS TO CONSIDER WHEN USING THIS APPROACH

Students who depend on memorised rules have to be helped to detach from them when they are not fit-for-purpose.

Solving equations can be associated with graphing so that the process and roots can be understood meaningfully, for example: when the height of a rectangle is 3 units less than the width, when is the area 0?

## Construction tasks

### EXAMPLE

Program a robot to draw regular polygons for which the side length and internal angle can be varied.

Construct a quadratic graph that has a minimum at a given point but whose gradients can be varied.

### ASSUMPTIONS ABOUT PRIOR EXPERIENCE

Availability of tools such as physical materials, ICT, and time to work on an extended task

### POSSIBILITIES OF THIS APPROACH

In these tasks algebra is a tool for controlling computer actions, giving instructions, expressing design features, and describing output variables. Creating an algorithm for carrying out a computation or transformation requires an understanding of relations between variables, unambiguous conventional notation[2], knowing how input relates to output and in what order operations have to be done.

Objects can have variable qualities, or be subject to some constraints such as having to conform to certain relationships.

### LIMITATIONS OF APPROACH

Energy can be lost in the making phase, and plenty of time is needed for trial and adjustment. The task needs to be designed so that students do not stick with *ad hoc* methods but need symbolic tools.

### SPECIAL CASES AND LINKS TO CONSIDER WHEN USING THIS APPROACH

Students need to: see potential cross-curricular and workplace implications; identify independent and dependent variables; use nested procedures and recursive methods; and to use iterative methods approaching a given level of accuracy.

[2] Note that software is becoming able to read ambiguous and unconventional notations and ask the user to choose between alternative precise versions.

# Summary

In this chapter we have presented many of the common difficulties students have with algebra, and explained why these occur and how teachers and researchers try to make a difference. We reviewed the research about various teaching and learning approaches and point out that to some extent these depend on curriculum aims and the quality of the teaching. However, what-ever the aims it is very common for students to be turned off mathematics if algebra teaching is mainly about manipulations which have no apparent purpose, in or out of mathematics. We have summarised and evaluated several approaches which can be combined to provide pathways for students to move from understanding relations between quantities (expressed as numbers) to the higher school uses of algebraic reasoning. In Chapter 8 we extend these ideas to think about solving equations and understanding more about functions.

# Where additional evidence is needed

Practitioner knowledge and experience can provide practical and contextual understanding of the problems of teaching. Connections between research and practice can be achieved through systematic development and evaluation. For example, three strategies can enrich teaching in a school and knowledge in the field:

- Construct and evaluate a map of how students' understanding and use of algebra can develop in one school, given the tasks and teaching approaches used.
- Two colleagues who teach parallel groups can co-design two teaching approaches for the 'same' algebra content. Give students pre and post-tests to diagnose their understanding and evaluate the outcomes of the different approaches.
- Students can be given similar tasks to those in the research papers mentioned in this chapter. Findings can be compared to those in the study to identify how pedagogy might influence the results.

## Key readings

Bednarz, N., Kieran C., and Lee, L. (1996). Approaches to algebra: Perspectives for research on teaching. In N. Bednarz, C. Kieran, and L. Lee (Eds), *Approaches to algebra: perspectives for research on teaching* (pp. 3–14). Dordrecht: Kluwer.

This introductory chapter to a collection of papers summarises many of the main problems students' have with algebra and the main components of learning algebra.

Jones, I. (2008). A diagrammatic view of the equals sign: Arithmetical equivalence as a means not an end. *Research in Mathematics Education*, 10, 119–133.

This paper presents a way to handle equivalence and substitution by physically moving numbers and expressions.

Kieran, C. (1992). The learning and teaching of algebra. In D. A. Grouws (Ed.), *Handbook of research on mathematics teaching and learning* (pp. 390–419). New York: Macmillan.

This chapter is an overview of the research about learning algebra and gives plenty of examples to illustrate its main points.

Küchemann, D. (1981). Algebra. In K. Hart (Ed.), *Children's understanding of mathematics 11–16* (pp. 102–119). London: Murray.

Children's understanding of algebra has not progressed much since this chapter was written. We have summarised its findings in our chapter, but Küchemann's work contains many interesting questions which can be used to assess students' understanding.

CHAPTER 3

# Ratio and proportional reasoning[1]

## Introduction

This chapter uses research about students' understanding of ratio and proportional reasoning to shed light on difficulties which seriously hinder success across mathematics. First we describe the nature and scope of the concepts, drawing particularly from a paper on rational number by Nunes and Bryant (2009). There is extensive research on younger children's developing understanding of these ideas – too much to be presented here in any detail – but we can indicate the source of many problems. We outline the relevant knowledge students might bring with them from their early schooling and describe what is known about ratio and proportional reasoning in later school mathematics. Some related concepts appear everywhere in the curriculum but are often treated implicitly, such as when learning about measure (see Chapter 4). We examine the strengths and limitations of various teaching approaches and make suggestions about how teachers might work with colleagues on teaching ratio and proportional reasoning.

---

[1] The preparation of this chapter was aided considerably by the Rational Number Reasoning Database hosted by NC State University, USA.

# The nature of ratio and proportional reasoning

## Definitions

Mathematical dictionaries tend to describe ratio as the quotient of two numbers and proportion as a comparison of ratios, with 'proportional to' indicating equality of ratios. These formal definitions give access to neither all the meanings which arise in the school curriculum, nor to the complex understandings that experienced mathematicians have of these ideas. Mathematicians who are learning to be teachers find it very hard to construct definitions that encompass their experience, yet they have no trouble recognising when to use these words and their associated techniques in a variety of mathematical and other contexts. The elusive quality of the concepts leads some textbooks to define the ideas in a very narrow sense. For example, for a while the National Numeracy Strategy in the UK, as several non-mathematical dictionaries, defined proportion as the comparison of part to whole, and ratio as part to part. This distinction leaves us with no way to describe the similarity of shapes, where not only are corresponding sides in equal ratios, but also the corresponding internal ratios of each shape are equal. These definitions also complicate learners' understanding of the constant relation between independent and dependent variables in proportional relations, indicated by $k$ in $y = kx$. The expert sense of ratio and proportion cannot be fully pinned down in school by definitions, so in this chapter we are going to look at the roots of understanding. Meaning accumulates through use, over time, in many different mathematical, everyday, and scientific contexts. This is the strongest recommendation to emerge from this chapter – the need to provide students with repeated and varied experiences, over time, so that multiple occurrences of the words and the associated ideas and methods can be met, used, and connected.

## Multiplicative relationship

The fundamental concept behind ratio and proportional reasoning (RPR) is the multiplicative relationship in which quantities, whether discrete or continuous, are compared using scalar multipliers. For example, we can say that one bucket holds three times what another bucket holds, or a sweater shrinks to four-fifths of its former size, or that I have twice as many sweets as you do. Note, however, that these three comparisons are very different in quality: the first is about capacity and the last about counting, but the second could be about tape-measured size, or cross-sectional area. Whatever the qualities, to make these statements we do not need to know anything about the actual sizes or quantities, we just need to be able to compare them. This is similar to saying 'the gradient of this linear function is 2'; we are not talking about the gradient between two particular points, but a relationship between any two points. The ratio between quantities is a comparison using the multiplicative relation, similar to the use of 'difference' to compare quantities in the additive relation. However there is a crucial difference between 'difference' and 'ratio' which makes ratio significantly harder – its numerical form does not represent a magnitude in the same units as those it is comparing.

 www.nuffieldfoundation.org/ratio-1

 www.nuffieldfoundation.org/ratio-6

If I compare two distances in miles additively, the difference is expressed in miles, but if I compare them multiplicatively, the ratio is expressed as either a pair of numbers or their quotient—the multiplier. For example, the comparison between my 12 minute walk to the shops and your 8 minute walk is 4 minutes using the additive difference, but the comparison using the multiplicative relation is that your walk is 2/3 the duration of mine. When the ratio can be counted, such as counting out three sweets for me to every two for you when sharing sweets in the ratio 3:2, the abstract nature of ratio is hidden in the concrete models and actions, but when it cannot be counted, such as in our journeys to the shops, the ratio has significant imaginary qualities. To make sense of two of your minutes being equivalent to three of mine (both a quarter of our walks), we have to imagine distances and speeds and starting times.

## Ratio as number

A scalar multiplier definition of ratio claims that it is the number achieved by dividing two numbers. If sharing sweets in the ratio 3:2 we could call the ratio of mine to yours after any completed round of the share-out as $\frac{3}{2}$ as in 'I have $\frac{3}{2}$ as many as you do' and the ratio of yours to mine $\frac{2}{3}$. But we know that comparisons between three or more quantities can also be expressed as ratios using the $a : b : c \ldots$ notation. If three waitresses earn different amounts per hour, we can express the ratio of their pay for the same hours worked as, for example, $60 : 69 : 63$. This simplifies to $20 : 23 : 21$. To use a scalar multiplier definition for ratio would be difficult unless we fixed one of them as a base unit, while a part-to-part definition would lead us to ask 'part of what'?

In practice the usual approach is to act according to the context. For example, if we needed to share out the Christmas bonus to the waitresses in these same ratios we could construct a 'unit' with which to measure. In this case, the unit would be $b/64$ where $b$ is the total bonus amount, and the waitresses get 20, 23, and 21 units respectively. 'Part' is a common term for such a unit. When mixing paint we say 'one part white to four parts blue', for example. Expressing these as fractions is complicated because we have to be clear about what is being compared to what. White makes $\frac{1}{5}$ of the whole, and there is $\frac{1}{4}$ as much white as there is blue. Blue makes $\frac{4}{5}$ of the whole, and there is 4 times as much blue as white. The multiplicative relationship between white and blue expressed as a number is $\frac{1}{4}$ or 4. Visualising the whole as being five of something is not obvious, but comparing white to blue makes more sense because we can visualise using equal-sized containers to pour from: one full of white and four full of blue (Figure 3.1). For every one of white we use four of blue.

**Figure 3.1**

The phrase 'for every' is a crucial way of quantifying the actions that lead to the final quantities being expressed in a certain ratio. Expressing visualisable actions as fractions can enable useful reciprocal and additive relations to be deduced, such as $\frac{1}{4}$ and 4 representing inverse relations, and $\frac{1}{5} + \frac{4}{5} = 1$ in the example above – all expressible through paint-mixing actions.

Ratio expressed as a fraction, an ordered pair, is a powerful link with the idea of measurement (see Chapter 4) and probability (see Chapter 7). The link with measurement is evident in our use of radians to measure angle. Angle is measured with the ratio of circumference to turn in a unit circle, so three radians is $\frac{3}{2\pi}$ of a full turn, and the associated arc is $\frac{3}{2\pi}$ of any circumference. A similar idea appears in the work of Confrey (1995) whose students called ratio a 'little recipe' to encapsulate its meaning as a measuring unit for a mixture. The little recipe idea is like this: for a small number of biscuits we need two spoons of flour to one of butter and one of sugar. For larger numbers we need multiples of these 'recipes': for three times as many biscuits we need three lots of 'two spoons of flour, one of butter and one of sugar'. For ten times we need ten lots of the little recipe. The distributive law allows us to gross up all the amounts of flour, butter, and sugar.

Turning ratios into single numbers, rather than the ordered pairs we use for fractions, is fraught with difficulty. In the paint mix just described, the ratio of white to blue expressed as in decimal notation is 0.25; the ratio of blue to white is 4; but the ratio of white to the final mix is 0.2, and of blue to the final mix is 0.8. The fraction notation carries more meaning than decimals, particularly if students understand it to represent division, or 'so much per so much', but we expect older students to understand gradients and trigonometric ratios expressed as single numbers using decimals, (e.g. sine 30° = 0.5) while understanding them to be ratios (as multipliers). For example, many teachers introduce sine as 'the number we have to multiply the radius by to get the height of a point on a circle'. In this diagram AB would be seen as a particular multiple of OA for the angle AOB (Figure 3.2). To understand this, students have to be fluent with the concept of multiples less than one.

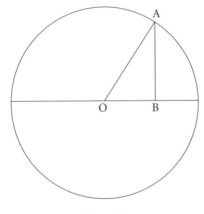

**Figure 3.2**

In geometry, ratio and reciprocal ratio are the multipliers with which we relate the lengths of two similar shapes. It is more usual in everyday life to say 'this shape is two-thirds the size of that one' than 'the ratio of lengths of this shape to that is 2 to 3'.

These connections between ratio, rational number, and fractions are presented in several ways in textbooks (Clark *et al.*, 2003):

- *ratio as subset of fractions*: fractions are any numerator divided by any denominator; ratios are the simplified versions of these called 'rational numbers';
- *fractions as subset of ratio*: ratio includes all multiplicative comparisons between quantities, and fractions represent ratios of part to whole, 'so much per so much';
- *ratio/fraction as separate*: ratio being part-part and fractions being part-whole;
- *overlapping sets*: both fractions and ratios have meanings which are not common, such as fractions representing a particular part of a whole, and ratios representing a unit of measure, such as density;
- *identical sets*: anything that can be represented as a fraction is also a ratio and *vice versa*.

We have shown already that some of these distinctions are limiting or spurious. Clark *et al.* observe that the 'overlapping sets' model is closest to the mathematical meaning, but can appear to be ambiguous when used in a school curriculum. Our own exploration of textbooks from several countries indicates a further difference: some using a fractional notation to express part-part ratios, one part being expressed as a fraction of the other part, so that a ratio of *a:b* can be written as 'the first quantity is $\frac{a}{b}$ of the second quantity'. Most textbooks avoid this latter use of ratio notation.

## Measurement and units

Ratio is intimately tied into our understanding of measurement because it is while measuring that students meet rational numbers. For example, if I am trying to fill a bigger mug from a smaller one I may want to say that the larger one is three-and-a-half times the smaller, or that two of the bigger hold the same amount as seven of the smaller. When measuring, quantities are compared using a common unit, of which they are both multiples, such as centimetres or litres (Behr *et al.*, 1992).

 www.nuffieldfoundation.org/ratio-1

 www.nuffieldfoundation.org/ratio-2

Experience with choosing suitable units for measuring provides a basis for understanding the part-whole relation, for example $\frac{1}{10}$ can signify different quantities depending on what the whole is. The metric systems lend themselves to using a zoom metaphor to decide the appropriate size of unit, so the meaning of ratio which is embedded in scaling lengths is also relevant.

## Rates

Seeing ratio as a fraction works if you are comparing parts of a single whole, or parts to a whole. The embedded actions on which this is based are partitioning or sharing. For example, I share one cake between six people means they each have a sixth; and if I share 24 sweets among six people they each have 4, which could be written as $\frac{24}{6}$, and is a sixth of the whole collection. But ratio is also important in comparing magnitudes that are similar but belong to different wholes; the embedded actions here are enlargement, stretching, or shrinking to match. For example, a well-known toy store sells various sizes of cartoon character dolls whose heights, head circumferences, arm lengths, and so on can be compared. Ratio is the way to do this, although there is no partitioning or sharing going on, nor any mixture for which there can be a mini-version. Within mathematics, the relation between the lengths of two similar geometrical shapes is of this kind.

A further type of comparison is between different quantities (Ben-Chaim et al., 1998). Ratios expressed as numbers can be scalar relations between quantities of the same stuff or compound measure relating two kinds of stuff, such as one egg per tablespoon of flour when making pancakes. A key word in this second meaning is 'per' which means we are always comparing the first quantity to a single unit of the second stuff. An oversimplification which can be found in some textbooks is that quantities to be compared 'have to be measured in the same units'. There is no mathematical reason for this, and if it is adopted it means that measures such as miles per gallon, or price per packet, seeds per gerbil, which compare different kinds of measure, produce a unit which has to be called 'rates' rather than ratios. These are also sometimes known as 'intensive quantities' (Nunes et al., 2003). The embedded action for understanding this kind of ratio is that other amounts can be made by self-similar additive or multiplicative replication of a mini-version, as in the biscuit recipe we gave earlier. Whatever the final quantity of gallons, packets, gerbils, or biscuits the relation between that and the miles, price, seeds, and ingredients involved is fixed.

Mathematically there are two things going on here: the notion of 'rate' as the gradient of a relationship, the rate of change, between two variables; and the notion of a mini-version which acts as both an additive and multiplicative unit.

www.nuffieldfoundation.org/ratio-2

Distinguishing rates from ratios might provide a bridge to students' understanding that multiplicative comparisons can be expressed as single numbers. These single numbers can be used to reconstruct specific cases of miles and gallons, seeds and gerbils, and so on. A mathematical example of this is how trigonometric ratios expressed as single numbers can be used to reconstruct side lengths. However, to separate rates from ratios for teaching purposes by over-definition is artificial and gives everyone, including teachers, something else to remember and get confused about. Rather than debate a dividing line between 'ratio' and 'rate' it is more important to understand that ratio and proportional reasoning are complex concepts which arise slightly differently throughout both mathematical and outside contexts. Confrey and Smith (1994) point out further that eventually we would like students to understand that rates can themselves vary, being themselves functions of the independent variable, as with the changing gradients of non-linear functions.

The previous sections highlight the multiple meanings and contexts for RPR, and a full understanding of RPR can only develop over time through all the above usages.

# Early school experience

## Early understanding of multiplication and division

From the above description it is clear that students' understanding of multiplication when they enter adolescence is important. Most children will understand multiplication as repeated addition, seeing it as a quick and efficient method for counting in groups of objects or an array. Many will know multiplication facts which they can apply in an abstract way to numbers, and some will understand division and multiplication as inverses of each other, though for many that will only be in the context of discrete objects or abstract rules for operating on numbers, rather than them having meaning for continuous quantities or indicating a particular relation (Booth, 1981). They are unlikely to recognise and express all the different transformations in which they might meet multiplicative relations (see Chapter 2) and might find it hard to decide when multiplication or division are appropriate operations in problems. They may have an understanding of scaling and enlargement, but might not associate with the word 'multiply'. Their understanding of scaling might be limited to doubling and halving, and they may not know what changes and what stays the same after enlargement.

Young children's understanding of division might be limited to situations which can be modelled by sharing. Sharing situations might be limited to part-whole situations in which children want the dividend to be greater than the divisor for division calculations to work (Behr *et al.*, 1992; Fischbein *et al.*, 1985). Children who think about sharing as a many-to-one and one-to-many correspondence, such as in cutting cake slices for several people, understand it when the dividend is smaller than the divisor (Nunes and Bryant, 2009). Those who have developed this many-to-one and one-to-many idea are more likely to have a sense of the inverse relation between, say, number of children and portion size of cake, as well as the direct relation between number of parts and number of children who can have cake. When filling in missing numbers in proportionality problems, about three-quarters of students in the 10 to 12 age range are not fluent with inverse problems such as: $\frac{2}{5} = \frac{10}{x}$ (Nunes and Bryant, 2009). If written as upright fractions, students might notice that they could use equivalent fractions where possible but only if they have a good knowledge of these.

To understand ratio students need to understand division as a *comparison* between quantities of whatever relative size. Inhelder and Piaget (1958) identify comparisons of quantities as the start of proportional understanding, but most students will not have a strong understanding of RPR before secondary school, even if they can successfully compare quantities.

## Measurement and RPR

A crucial step towards multiplicative reasoning is understanding iterated units (Outhred and Mitchelmore, 2000). Although students aged 9–13 often have proficient linear measuring skills, they do not necessarily understand the underlying principle of an iterated unit which covers a given quantity. Barrett and Clements examined the understanding of length and the development of measuring strategies (2003) and found that children initially use an additive operation to measure length. As they develop more abstract strategies to deal with length, the additive idea becomes integrated into a multiplicative and iterative scheme which might contribute to a basis for proportional reasoning.

The inverse relationship between the size of unit and the number of units required, for example to cover a given area, is a difficult concept (Hiebert, 1981). For example, dividing by 0.2 gives a result 100 times greater than dividing by 20.

Measuring area is very different from measuring length. If area has been used as a model for ratio, for example by shading a rectangle in two colours to represent boys and girls in the class, the associated understandings may be limited to counting squares. The use of shapes to represent the 'whole' can confuse partitioning into equal quantities with partitioning into congruent shapes. For example, when drawing quarters of squares students usually produce four congruent shapes (squares or strips) rather than non-congruent quarters of equal area (Figure 3.3).

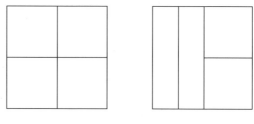

**Figure 3.3**

This suggests that area models do not necessarily convey the sense of quantity that is necessary for ratio.

## Ratio, rates, and linear predictions

Students' quantitative understanding may include being able to use intuitive and *ad hoc* methods in ratio situations when they understand the fundamental relationship in contexts (Ben-Chaim *et al.*, 1998; Nunes *et al.*, 1993). Some of this reasoning is multiplicative; for example, they might have some idea of rates such as speed or unit price. Knowledge of rates might come from cross-curricular work and they may not associate it with division. They may be able to use rates to calculate values through multiplication, using either known facts or calculators. Many 11-year-olds can make correct linear predictions which could provide a basis for developing RPR, such as knowing that if two metres of cord costs 80p then ten metres will cost £4 (Nunes and Bryant, 2009).

# Summary

**TABLE 3.1 Summarises the experiences students may have had in their first six years of school on which to build an understanding of RPR.**

| RPR knowledge at about 11 years | Potential | Limitations |
|---|---|---|
| **Active experiences** | | |
| Comparing quantities | Early school and out-of-school experience. | It is easier to compare using 'difference' rather than ratio. |
| Mixtures | Model for RPR; draws on familiar actions; can use mini-versions to build up bigger quantities. | Can be misinterpreted as additive. Confusion between part-part and part-whole fractions. |
| 'For every...' 'For each...' 'So much per so much' | Phrases which convey proportionality, rate, and correspondence. | Overuse can limit recognition of RPR in other situations. |
| Scaling | Embodies RPR; gives meaning of multiplication and reciprocals. Connects to measuring. | Not connected easily to counting, repeated addition, or correspondence. |
| Doubling and halving | Connects to scaling and reciprocals. | Does not mean that students understand multiplication. |
| Rates | Out-of-school knowledge, e.g. prices, fuel consumption. | Reduces ratio to one number; need to connect this to other notations. |
| Linear reasoning | Familiar from outside experiences. Relates to proportion, equivalent fractions, graphs of constant rate. | Can be misapplied; need to know what is *not* proportional. |
| **Calculations** | | |
| Multiplication with whole numbers | Multiplication facts speed calculation; to understand division as inverse and the consequent need for non-integers. | Can be limited to a repeated addition model; division limited to known cases; fraction notation for division is not usually used in this context. |

(continued)

**TABLE 3.1 (continued)**

| RPR knowledge at about 11 years | Potential | Limitations |
|---|---|---|
| **Active experiences** | | |
| Multiplication with continuous number | Expands meaning of multiplication to scaling; leads to needing decimals and fractions. | Can be limited to techniques without understanding. |
| Division as inverse of multiplication | Useful if 'undoing' arrays and in counting and scaling contexts. | Algorithms for division do not explain this meaning. |
| Division as sharing | Helpful in developing correspondence and constructing fractions. | Can limit meaning to part-whole contexts. |
| Division expressed as fraction | Useful if both quantities are integers; encapsulates the meaning of division; easy to use reciprocal; can express equality of ratios as $\frac{a}{b} = \frac{c}{d}$ and use equivalence. | Tendency to reduce fractions to rules for procedures, when often understanding equivalence is enough; spatial part-whole models tend to dominate. |
| **Definitions** | | |
| Ratio defined as part to part | Useful when comparing parts of an identifiable whole. | Not useful when comparing different objects. |
| Proportion defined as part to whole | Useful when thinking about a clear 'whole' and its components, e.g. mixtures. | Not helpful for different objects. Not helpful for fractions >1 and percentages >100. |
| Proportion as equality of ratios | Mathematically correct and complete definition. Relates to graphs of $y = kx$. | Need to manage several variables. Tendency to resort to, and misuse, methods. |

# Students' understanding

Most of the information about the growth of understanding of RPR is found in studies which either assume, or involve, a particular curriculum or teaching approach. Students' ideas are the products of active minds, trying to make sense of their limited experiences. The teacher's task is hence to think in terms of what additional and new experiences they need rather than re-teaching using past approaches.

## Roots of ratio

Many out-of-school experiences involve RPR, and understanding grows throughout childhood, entwined with multiplication and division. Formalisation of these ideas that ties students to particular notations and methods does not always draw on the experiences of comparing quantities, both discrete and continuous, that are central to understanding RPR. Young children can compare quantities without quantifying them, knowing about 'bigger', 'smaller', and 'the same', and also knowing that 'so many of this equal so many of that'. They can make judgements about quantities and qualities and rates, for example speed, crowdedness, or price per unit. In some teaching studies there is evidence that they can do this with continuous quantities, such as stronger or weaker fruit drinks, and also discrete quantities, such as collections of different coloured ping-pong balls (Confrey, 1995; Kaput and Maxwell-West, 1994).

One meaning of fractions is the part-whole relation between the numerator and the denominator which can be embedded in how we read the fraction notation out loud (Armstrong and Larson, 1995). Spatial representations, or representations based on sharing food, can generate conceptions of fractions as parts of a whole but do not enable the more powerful conceptions of fractions as entities, such as being a value on the real number-line (Tzur, 1999). Toluk and Middleton (2003) observed that if part-whole is the only way students think about fractions, there is a resistance to considering the fraction as a quotient. A further meaning is as an operator, as when finding a fraction of another number. These considerations point to the importance of understanding how the notation, the spoken word for the fraction, and the underlying quotient, number, or operation are related.

## Why RPR is so hard

Cramer and Post (1993) identified the range of successful strategies used by 12 to 14-year-olds who were given three different kinds of PR problem: missing value; numerical comparison of ratios; and prediction about changes in quantities.

In missing value situations the student has to decide what to compare with what, and whether this is an inverse or direct multiplicative relation. In each missing value problem there are potentially four different ways to set up the relation $\frac{a}{b} = \frac{c}{d}$. The questions they used were based on work of Karplus *et al.* (1974) whose task was about the heights of two figures:

Mr Tall measures 6 buttons; Mr Small is 4 buttons or 6 paperclips in height. How high is Mr Tall in paperclips?

Students could compare buttons to buttons and paperclips to paperclips, or could compare Mr Tall's height measured two ways to Mr Small's height measured two ways. The number six appears twice, which could be confusing and also masks whether students are using 'within'- or 'between'-case comparisons. Use of formal algebraic methods would give $6/4 = x/6$ or $4/6 = 6/x$. The difficulty of such problems varies with the relations between numbers.

The students used four main groups of strategies: identifying a unit rate; identifying scale factors; matching equivalent fractions; and cross-multiplying. The scale factor approach was only used when it could be expressed as an integer; equivalent fractions were used more by the older students; and cross-product was only used by older students and was prone to be misapplied. Unit rate was used more by younger students. Choices are clearly influenced by teaching because cross-product is a procedure that does not arise directly from meaning, but from condensing several actions. The original research by Karplus and his colleagues for the missing number problem also shows a variety of additive strategies leading to wrong answers.

Understanding proportionality in the way it is understood by mathematicians includes handling situations involving four variables when $\frac{a}{b} = \frac{c}{d}$. Full understanding includes knowing that an increase in $a$ can be compensated by decreasing $d$, or increasing $b$, or increasing $c$ (Noelting, 1980a, 1980b).

There are, therefore, many intermediate stages towards full understanding in which students might see what to do in one situation but not in another, even when they are structurally identical (Tourniaire and Pulos, 1985). Any extra hurdles, such as non-integer ratios, co-prime denominators, unfamiliar

www.nuffieldfoundation.
org/ratio-4

 www.nuffieldfoundation.
org/ratio-5

wording or contexts, can lead students to resort to additive strategies (such as building up and comparing) instead of multiplicative (Hart, 1981). Even when students are fluent in techniques they may have been taught to solve particular problem formats, such as cross-multiplying, they are likely to revert to *ad hoc* methods when the format alters, and also to misuse the method in inappropriate contexts. Knowing how to choose variables to control and deal with one change at a time helps, because it reduces the need to manage several changes at once. Students who think about the meaning of a problem before working on it do better than others (Karplus *et al.*, 1983).

 www.nuffieldfoundation.org/ratio-3

In mixture problems, students have to focus on the ratios involved, that is the relations between the ingredients. Counting is not an option because of the continuous quantities involved. Representation with diagrams is hard because a new kind of object has to be depicted. Eventually mathematics students need to be able to provide their own representations, but provision of physical manipulatives and suitable representations can help students work with such problems (Kieren and Southwell, 1979). With all manipulatives the underlying purpose needs to be to help students comprehend the problem structure and recognise it in future. Some mathematical representations can lead to unintended interpretations, and diagrams that present the components separately can lead to students thinking about difference rather than ratio unless the context is familiar enough to over-ride this potential confusion.

For example, consider problems that ask students to compare orange juice mixtures where the grey is supposed to be the juice and the white the water (Figure 3.4). Because the components are represented separately, it is tempting to compare heights of the grey sections by subtraction to describe the differences in the mixtures instead of comparing grey to white.

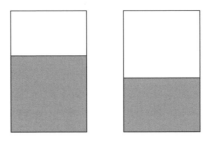

**Figure 3.4**

## Additive to multiplicative methods

Students have difficulties with situations where the scale or ratio is not an integer, or which cannot be solved with doubling or halving, or where numbers are co-prime. This suggests that they are limited to additive or building-up methods. Although responses to RPR tasks are dominated by additive methods, Hino (2002) found that when using a unit that gives insight into a situation, rather than just as a label, students can use multiplication in novel situations. We have already mentioned this approach reported by Confrey (1995) in which mini-versions of the whole mixture are used as units which can bridge between additive and multiplicative methods. Kaput and Maxwell-West (1994) provided software which allowed 11-year-old students to chunk and handle relations in ways not possible without help. In their study, students could create onscreen icons with which to deal with proportional situations, and use the icons to generate their own correct solution methods rather than setting up and solving an equation. They found in case studies that their use of the phrases 'for each' and 'for every' were critical in supporting understanding.

There are strong reasons for setting up ratios as fractions so that equivalence can be used as an effective multiplicative strategy. Research reported by Tourniaire and Pulos (1985) shows that the ability to select an appropriate method, either constructing a scale factor or constructing a rate, is more advanced than the ability to make the fractions equivalent. In other words, using equivalent fractions to find missing values is easier than identifying the constant of proportionality.

When students do understand and choose to use multiplicative strategies, their choice is influenced by the context (Lamon, 1993, 1996) and problems might be tackled using within-case or between-case ratios, or as a rate to compare different variables, or as a scalar multiplier to compare similar mixes. In the Mr Tall and Mr Small problem this means that it could be tackled by comparing ratios of buttons to paperclips for Mr Tall and for Mr Small (within-case), or by comparing ratios of buttons to ratios of paperclips (between-case), or by setting up a rate 'so many paperclips per so many buttons', or as a scalar multiplier: 'Mr Tall measured in buttons is k times the height of Mr Small'. Lamon (1996) categorised students' methods as non-constructive – such as using visual, additive, or pattern-based approaches – and constructive – such as qualitative and quantitative proportional reasoning. She explored the reasoning of a cross-section of 11-year-olds who had not yet been taught any formal ratio methods and found that many used informal language for ratios. Correspondence problems, many-to-one and one-to-many, generated most proportional reasoning but, as with other studies, students resorted to additive methods when the scale factor was not an integer.

In many studies, problems about the ratio 1:2, or problems with numerically related ratios such as 1:2 and 1:4, turn out to be the easiest to solve (Hart, 1981; Noelting, 1980a, 1980b). In these examples it would be hard to know if students were using the multiplicative relation or seeing doubling as a special case of addition. Research shows an increase of difficulty between 1:2 and other 1:n ratios, with another increase in difficulty $n:m$ $(n \neq 1)$ (Tourniaire and Pulos, 1985). Clearly a combined approach is needed: both understanding the relations and also feeling able to do the appropriate calculations. It is not just a matter of whether to do them by hand or by calculator, it is also about knowing what operations to do.

 www.nuffieldfoundation.
org/ratio-1

 www.nuffieldfoundation.
org/ratio-4

## Division

When faced with division calculations in contexts, that is without being told when to perform a particular algorithm, students use a range of strategies including: skip counting; repeated addition or subtraction; seeing the dividend as chunks of something ('so many of …'); using known or derived facts; trial and adjustment; and modelling based on context. Written algorithms are based on repeated subtraction of multiples from chunks of the dividend and do not easily match students' situated attempts to handle the quantities involved in the contexts.

Quotative situations in which the number of groups of equal size has to be calculated are harder for children than partitive, in which the dividend has to be split into a given number of parts. For example, it is harder to deal with questions like 'If we are sharing out three cakes, how many children can be given a quarter of a cake each?' than 'If we are sharing out three cakes among twelve children, how much can they have each?' In the second question the answer is a portion of the given quantity; in the first, the answer is not in terms of the given quantity.

Division arises in many ways:

- as a component of the multiplicative relationship;
- as the inverse of multiplication achieved by reversing multiplication facts, usually associated with integers;
- as algorithms that combine chunking numbers, reversing multiplication facts, and repeated subtraction until zero, a remainder, or a decimal representation is achieved;
- in fraction notation where fractions are a meaningful outcome and representation of division, and cancelling by common factors simplifies the division;
- with non-integer answers and hence leading to fractional answers;
- as the result of partitioning continuous or discrete quantities, or unit quantities, or numbers, into *n* parts, expressed as fractions or decimals;
- as the number of parts achieved for a given portion size (as fractions or decimals);
- as the 'undoing' of enlargement using a reciprocal scale factor.

Limited experience can lead students to understanding division only in certain domains such as inverting multiplication of integers, sharing integer quantities, assuming the dividend has to be larger than (or smaller than) the divisor, etc.

When young children are faced with the problem of finding out how many smaller vessels of water fit into a larger vessel, they typically count the actions and if this is an integer they can then compare quantities multiplicatively. If the answer is not exact they can eventually understand that a smaller 'measuring' cup might be used for both. This measure is a common factor of the capacities of both the original vessels. The most efficient such measure is the highest common factor (greatest common divisor). An alternative way to solve the problem is to find how many of the larger vessel and how many of the smaller vessel would be equivalent; the most efficient equivalent measure in this approach being the lowest common multiple. These experiences provide a basis for understanding that common factors and common multiples are tools for comparing quantities and numbers, and both methods lead to the expression of the ratio of the two original amounts. More explicitly, if five measures fill a smaller vessel and nine fill a larger vessel, the ratio of the capacities, smaller to larger, is 5:9. Also, nine of the smaller are equivalent to five of the larger, because $9 \times 5 = 5 \times 9$.

www.nuffieldfoundation.org/ratio-6

When fractions arise as a way to express rates, division is implied, but the unit is always a compound measure, such as pounds per kilo, or miles per litre, or mass per cubic centimetre. For example, suppose we weighed and measured a cuboid of copper and found that it weighed 45 grammes and was 5 cubic centimetres in volume. Setting this up as a fraction, $\frac{45}{5}$, and using the common factor 5 to simplify it to $\frac{9}{1}$, preserves the sense of ratio as 'so much per so much'. In fact the density of copper at room temperature is 8.92 grammes per cubic centimetre, but understanding how to use that statement depends on understanding first of all how such a number is related to the actual properties of a given case.

## Linear understanding

Once students accept multiplicative relations they are likely to make linear assumptions even when the situations they are working with have a constant term, are exponential, or involve powers of $x$ greater than one. For example they will think that if a function has value 10 when the independent variable is 2, it must take the value 20 when the variable is 4. Whether this is developmental or due to educational hierarchies has been investigated in some depth by De Bock and his colleagues (De Bock *et al.*, 1998, 2002). In area tasks they found a common belief that any 'increase' is taken to mean 'proportional increase'. This persisted despite formal learning and there was an intuitive or deliberate reliance on proportional reasoning. They warn that the presence of the linear response does not mean that students understand proportionality. Hadjidemetriou and Williams (2010), in a study of 425 15-year-olds using and applying graphs, found that a linear assumption was more likely to be applied erroneously by higher attaining students, maybe because they have adopted this as an all-purpose tool, while lower attaining students were more likely to be confused about the meaning of the graph. As De Bock *et al.*'s work suggests, one aspect of understanding proportional reasoning is knowing what situations are NOT proportional as well as what are, and the use of tasks which expose contradictions might be pedagogically valuable (Psycharis and Kynigos, 2004). Thus students should learn to discern structural similarities beyond perceptual similarities. Modestou and Gagatsis (2010) tested 562 12- to 14-year-olds to see if they could distinguish between linear and non-linear situations. Students who could do standard tasks well could not necessarily make the distinction. It is possible that the tendency to use linear assumptions inappropriately is because of their close relationship to additive models through repeated addition, and this may also contribute to their misuse.

## Progression

In our experience, the complexity of the notion of division as laid out above is a revelation for many specialist teachers of older students. Considerable time needs to be devoted to experiencing division in a variety of ways, to extend students' concept images and available methods, before embarking on organised work about ratio. We can also say with some confidence that if students do not have any understanding of multiplication other than repeated addition, they are unlikely to understand ratio and may cling unhelpfully to a quasi-multiplicative models based on incremental steps. If they do not have the knowledge and tools to express multiplicative comparisons through constructing and simplifying fractions, and as decimals through calculation, they are likely to find RPR very frustrating. Whether students learn about calculations through working with RPR, or learn about RPR through doing calculations, or learn about both by working on tasks that develop meaning in contexts, is not fully understood, but a combination of all of these approaches over several years is the likeliest candidate for 'best' method, since that is how successful mathematics students report their learning, that is by sorting out random multiple meanings over time.

One of us recently observed a teacher ask a class of 12-year-olds what they knew about ratio. One student said 'it is something to do with division but

I cannot explain it very well' and the teacher replied 'well neither can I', which seems to be a reasonable answer given the complexities above.

## Teaching approaches

There is some limited evidence for the success of various teaching methods. For example, Lachance and Confrey (2002) and Moss and Case (1999) found that children who had previously worked with measurement in contexts then learnt about decimals and fractions more easily than expected. It could be worthwhile revisiting this connection in early adolescence and extending it to comparing lengths multiplicatively. Freudenthal argues for an approach which keeps cycling between formal, informal, and insightful experiences (1983, p. 209). He concluded that learning ratio and proportionality 'can be achieved [...] by returning again and again during the process of algorithmisation and automatisation, and even afterwards where it fits, to the sources of insight' (p. 209).

All teaching methods which have been reported in research show *some* effect because students have been taught something relevant, but in a comparison of such studies it was found that the length of time of instruction was also related to success (Confrey, 1995). In Japan it is typical to spend several weeks at the start of high school studying proportionality (and non-proportionality) as a central mathematical idea (Cave, 2007). Confrey developed RPR with students between 9 and 11-years-old (1995) over a few years. She saw time as a critical feature of her work, so that she and the students could build a shared repertoire of ideas, models, ways of talking, and past experiences to refer to. She started with scaling recipes for different numbers of people. We mentioned earlier in the chapter that this approach led students towards a strong understanding of RPR and also to the concept of distributivity.

Rate of change, in Confrey's view, has both additive and multiplicative meanings, and also an intermediate meaning of 'what has been added as a proportion of what was there before' such as with compound interest (Confrey and Smith, 1994). Students in her work made units by sharing, folding, magnifying, and also in splitting: the action of making multiple smaller versions of the original, such as in the 'little recipe'. These units could be added or multiplied with all their current properties retained, an idea close to similarity, proportion, and magnification.

Focusing on constructing such compound units holds the proportion constant and avoids problems with non-integer scalars. In several studies in which

www.nuffieldfoundation.org/ratio-1

a shift from additive to multiplicative reasoning can be seen, fluency with multiplication and division were problematic for many students (Kurtz and Karplus, 1979; Mitchelmore *et al.*, 2007). It is worth asking whether calculation methods can be bypassed with calculators or whether this would hide direct experience of the associated structures. For example, the gradient of a straight line can be expressed as 'goes up 3 units for every 2 units along' rather than having to express it as $\frac{3}{2}$, which is known to provide a technical hurdle. Students who fully understand fractions as representing a relation can see this expressed in $\frac{3}{2}$, but those who are tempted to use 1.5 or $1\frac{1}{2}$ can lose this sense of structure. Rates are an aspect of RPR where the most appropriate way to perform the division might be to set up a fraction and cancel common factors, where possible, to preserve the meaning, rather than performing a division algorithm automatically.

Ben Chaim and colleagues undertook a comparative teaching study with 215 12- and 13-year-olds. The experimental group was encouraged to construct their own procedures for solving RPR problems in context and were given no standard algorithms or processes to use (1998). He tested 'own procedure' and 'formal methods' groups afterwards with questions about rate, such as 'best buy' questions, missing values, comparing rates, and comparing density. A mixture of integer and non-integer questions was used. 'Own procedure' students did better in familiar contexts than 'formal methods' students. This confirmed that just because students have been taught methods does not mean they (a) use them or (b) know when to choose to use them or (c) feel able to tackle the associated arithmetic. The 'own methods' approach appeared to improve confidence in tackling RPR questions, ability to mathematise familiar situations, and ability to develop and use suitable representations. Ben Chaim suggests that the value of the 'own methods' approach is that it coordinates informal and school experience, but this raises the question of whether it did anything more than enable students to develop everyday *ad hoc* methods of dealing with RPR tasks, particularly as it did not affect their ability to deal with tricky cases.

An approach that requires the use of several different calculation methods, while maintaining the focus on ratio, is to pose a series of questions about a giant or a toy and their relationship to human dimensions. An early version of this was Streefland's use of an enlarged handprint (1984):

- estimate the giant's height;
- compare sizes of objects from the giant's world (sole of his shoe, newspaper, handkerchief) with those of the human world;
- compose cake recipes for giants;
- compare the steps (and speed) of giant and human;
- compare numbers and prices of things in both worlds;
- use number lines and pie charts to model invariance of ratios;
- develop a ratio table as a schematisation of ratio.

Ratio tables, such as those suggested by Streefland and colleagues, are a widely-used tool for keeping track of comparisons of multiplicative patterns. This table shows multiples of three and five; the relation between the rows is always 3:5.

| Length in human world | 6 | 9 | 12 | ... |
|---|---|---|---|---|
| Length in giant world | 10 | 15 | 20 | ... |

Such tables make the relationship between the rows accessible, and make it easier to handle two variables and their relationship at the same time. The table also shows the usefulness of using 3:5 as the replicable unit rather than expressing it as a decimal number. In textbooks from the Netherlands, ratio tables are used frequently to organise realistic numbers from authentic situations, that is not just small integers. For example, in an organisation it is recommended that the number of managers to the number of workers is a fixed ratio; give the number of managers for various numbers of workers:

| Number of managers | 75 | | | |
|---|---|---|---|---|
| Number of workers | 1200 | | | |

Since handling all the variables seems to be a major problem, we now look at an approach in which variation is deliberately controlled. Lobato and Thanheiser (2002) have used this effectively with a small group of fifteen students aged 15 to 18. They started with the need to understand ratio as an indirect measure of slopes of functions, and the recognition that students would typically confuse slope with height, points, and other absolute simple measures. They argued that if slope is only seen as a number then there is little sense of how it represents covariation. They let students use dynamic geometry to (i) isolate slope as an attribute by holding other variables constant; (ii) change the variables to see how they affect slope; (iii) explore how ratio-as-measure acts in relation to the function; and (iv) construct the slope ratio for themselves. By using variation themselves to explore the slope ratio, and constructing slopes that had varying features, students were able to relate the representations, the quantities, and the underlying variation. This study is one in which students' understanding of RPR is developed within a curriculum topic that is often taught after ratio. Planning includes developing RPR as well as the current topic. This could apply when teaching rates of interest, trigonometry, functions, calculus, vectors, and algebraic transformations. To use these as contexts for learning more about RPR fulfils the need for multiple experiences of RPR over time to extend and embed understanding, and also avoids trying to layer complex topics onto students' default methods of additive, build-up, or *ad hoc* reasoning. In this way more explicit attention can be extended to RPR without having to delay the introduction of more complex topics.

An alternative approach is to anticipate later needs for RPR in mathematics and construct explicit task sequences that lead students, over time, through a coherent programme of study. Carraher (1993) developed such a learning trajectory based on software that permits manipulation and comparison of the unknown lengths of two line segments together with parallel algebraic/ arithmetical and graphical representations of their equalities. His starting place assumes only that length is already understood:

1. *Common factors*: find divisors that create equal lengths on different line segments $a$ and $b$. For example, give two line segments of lengths 12 cm and 15 cm. The student divides the first by 4 and the second by 5 to get lengths of 3 cm.

2. *Common multiples:* find multiples that create equal segments from different lengths. For example, with the same two line segments as above, the first would be multiplied by 5 and the second by 4 to make 60 cm.

3. *Two-integer operations:* suppose $\frac{a}{n} = \frac{b}{m}$, then the line segment whose length is $\frac{a}{n}$ is multiplied by $m$ to show it is then equal to $b$. This gives a sense of the possibilities using inverse relations. For example, with the segments given above: $\frac{12}{4} = \frac{15}{5}$.

4. *Rational multipliers*: students find the multiplier $\frac{n}{m}$ that will make two lines equal in length. In the case we are considering this multiplier is $\frac{4}{5}$ so that $12 = \frac{4}{5} \times 15$.

5. *Rational divisors*: students find a rational number that when divided into one segment equals the second segment. By this time, students can be comparing the various roles that the numbers 4 and 5 are playing in their enquiry.

6. *Relative increase and decrease:* expressed as $a = b + \left(\frac{n-m}{m}\right)b$. In our example, the right-hand side is $15 + \left(\frac{4-5}{5}\right) \times 15$. This expresses the difference between the given lengths in terms of the multiplicative relation between them, thus laying the foundations that one length can be expressed in terms of another with no reference to external measuring units.

7. *Measurement:* $a$ can be 'measured' using $b$ or vice versa.

Later tasks in the trajectory lead students to find appropriate scalar multipliers to make $a = \left(\frac{n}{m}\right)b$ by iterating successive attempts to find $\frac{n}{m}$.

This task sequence focuses throughout on the meanings and uses of $\frac{n}{m}$. The strength of this approach is the constant comparison between a model which shows what is happening in lengths and the symbolic representations of the actions in number and algebra, and eventually with graphs. Not only is RPR modelled, but also the algebraic practice of 'doing the same thing to both sides of an equation' and the development of a linear relation. Another feature is that Carraher sustains a powerful, extendable, model through several iterations of task and meaning.

# Summary

The main features of learning RPR which have emerged in this review of research are as follows.

- RPR concepts appear everywhere in the curriculum but are often implicit.
- Non-integer ratios, co-prime denominators, and unfamiliar wording or contexts can lead students to resort to additive strategies.
- RPR is learnt over time in many different mathematical and scientific contexts.
- Concrete models can hide the abstract nature of RPR.
- Contexts where quantities cannot be counted encourage significant imaginary qualities of RPR.
- Students have to understand gradients and trigonometric ratios expressed as single numbers.
- Ratio is tied to measurement, comparing quantities using a common unit; young children can sometimes construct their own suitable units.
- Students are more likely to use unitary methods with contextual problems where they can 'design' the unit.
- Formal methods tend to be misapplied.

- Many problems with ratio are due to lack of facility with multiplication and division facts and methods, such as setting up an appropriate fraction.
- The relationship: $\frac{a}{b} = \frac{c}{d}$ has to be understood, including how one change affects other variables.
- A range of question types is necessary for full understanding.

So what are we left with for teaching advice? Nothing suggests that one way of teaching is better than another in itself for having some effect on RPR, but taking the studies together we conclude that learning RPR is a medium-term project and not something that can be 'sorted' in a few lessons. Growth of understanding has to connect to: knowledge of multiplicative relations; fractions and decimals; and outside and school-based knowledge of sharing, measuring, rates, and compound measures such as gradients. The few studies that report on use of taught procedures see them being used inappropriately as well as successfully, and not generally employed unless specifically requested. The students' understanding of the situation is more influential in choice of method than what has been taught.

We conclude that a whole-school approach is necessary to ensure that the complex ideas involved in RPR are developed coherently over time, following guidance arising from the research.

- Build on students' knowledge about contexts and relative quantities (that cannot be counted), while recognising that these maybe limited. Use these to identify variables and learn how to express relations.
- Build on students' capacity to design suitable units, such as 'little recipes', in contexts.
- Represent and carry out division in many ways, maintaining the connection between division and ratio by using simplified fraction notations.
- Represent ratio in a variety of ways: $a:b$, $\frac{a}{a+b}$, and as a decimal number. When expressed as a fraction, understand that this is a multiplier and not a part-whole representation.
- Using the simplest version of a mixture, that is the simplified ratio/fraction, as a unit to build with, and extending this idea to all relative measures.
- Use models (images, metaphors, diagrams, contexts) which are extendable over a considerable time: avoid models that might encourage over-simplification.
- Provide ways to 'hold' information about variables and their relations, for example through model use, ratio tables, fraction notation, graphs.
- Develop confidence with multiplication and division of non-integers.
- Use a range of types of problem that need proportional reasoning and cannot be done in other ways.

- Spend time on distinguishing linear (proportional) from non-linear situations (including linear functions which have non-zero constants).
- Investigate how changes in variables affect the relation: $\frac{a}{b} = \frac{c}{d}$.
- Systematically develop all these ideas over time, and give them time wherever they arise in later mathematical topics.

# Where additional evidence is needed

There is a major lack of knowledge of how understanding develops through school over time and across contexts.

The issues in the summary need to be discussed among colleagues to look for contrasts in how they are tackled by different teachers and in different years within the same school. The research reported in this chapter can be used to identify the possible advantages and limitations of different approaches. Tasks can be constructed to reveal how students' understandings develop through school.

The chapter gives information about typical student errors, but it could be the case that students of individual schools and teachers show different patterns of understanding. Set a range of question types to find out when students: use formal methods correctly/wrongly; use informal methods correctly/wrongly. Find out when students use multiplication and/or division as a first resort in unfamiliar contexts and how this relates to the relevant units.

The table of experiences up to age 11 can be used with feeder schools to find out whether there is agreement about curriculum order and difficulty at transition.

## Key readings

Behr, M., Harel, G., Post, T., and Lesh, R. (1992). Rational number, ratio, proportion. In D. A. Grouws (Ed.), *Handbook of research on mathematics teaching and learning* (pp. 296–333). New York: Macmillan.

This chapter summarises what is known about the component concepts of RPR and typical difficulties that learners experience.

Cramer, K. and Post, T. (1993). Connecting research to teaching proportional reasoning. *Mathematics Teacher*, 86(5), 404–407.

This short paper presents a range of question-types which can be used to structure a scheme of work, or to assess students' learning.

Nunes, T. and Bryant, P. (2009). Paper 3: Understanding rational numbers and intensive quantities. In T. Nunes, P. Bryant, and A. Watson, *Key understandings in mathematics learning: A report to the Nuffield Foundation.* http://www.nuffieldfoundation.org/ key-understandings-mathematics-learning.

This research synthesis contains many illustrations of how very young children understand quantities and hence provides a foundation for thinking about teaching older students.

CHAPTER 4

# Connecting measurement and decimals

## Introduction

The school mathematics curriculum for older students is commonly structured within a manageable number of headings; an oft-used configuration being number, algebra, geometry, and statistics. Whatever the chosen structure for the curriculum, there are always going to be mathematical topics that spread across and between such headings. One of these is measurement. This is because some aspects of measurement, such as measuring length or area, clearly relate to the geometric properties of shapes and hence could be listed under the geometry heading. At the same time, other aspects of measurement, such as time or money, are about number. In these ways, measurement is a topic that connects and enriches the two crucial mathematical domains of geometry and number and thence algebra, through measurement formulae and relations between quantities (Clements and Bright, 2003; Owens and Outhred, 2006; plus see Chapters 2 and 8). An example of the wider implications of connecting between, say, geometry and number is when students are learning to read and use maps. Here a critical factor for success is understanding scaling: something that is geometrical but which also entails both measurement and proportional reasoning plus, in all likelihood, decimals.

Measurement also links with probability and statistics, as probability can be thought of as a measure of uncertainty and the various statistics of data variation, such as standard deviation, can also be viewed as forms of measurement.

Indeed, what is sometimes called the 'uncertainty approach' to measurement takes the position that the information from a measurement 'only permits a statement of the dispersion of reasonable values' (ASE-Nuffield, 2010, p. 7). As such, all information from measurement is statistical, and all such information is approximate or uncertain (Bell, 1999). The link here with decimals is the idea of significant figures or digits as an indication of *mathematical accuracy* (usually taken as the number of significant decimal digits to the right of the decimal point) and *mathematical precision* (usually taken as the total number of significant digits), though note that both accuracy and precision are associated with slightly different meanings within measurement in science.

The implications for mathematics education that decimals arise when things are measured was noted in 1929 by Forno: 'Because of the extensive use of decimal fractions at the present day [i.e. 1929] in reading speedometers, cyclometers, statistical reports, etc., it is most important that pupils have clear ideas of the meaning of decimal fractions and be able to interpret them rightly' (Forno, 1929, p. 7). Nowadays, the use of decimal measures continues to grow, with students increasingly likely to encounter digital technologies such as sensors and data-loggers to gather and display measurements of motion, light intensity, temperature, and so on (JMC, 2011; Oldknow and Taylor, 1998). In many ways, measurement, invariably in decimal format, provides a fundamental link between mathematics and the sciences, art and music, the social sciences, and many other disciplines, and is pervasive in daily activities.

All these links between measurement and decimals, and the links between measurement and other components of the mathematics curriculum for older students, means that the placing of measurement in the mathematics curriculum can be problematic for curriculum designers and policy makers. In turn, it means that it is equally tricky for teachers to teach measurement and decimals in the most effective way. Perhaps as a result of its links, measurement can either seem all-pervasive within school mathematics, or, alternatively and probably more likely at secondary school level, sometimes appear as a somewhat neglected topic in comparison to more overt and apparently sophisticated components such as, say, algebra. As Ainley concluded (1991, p. 76) 'There is mathematics in measurement; but it does not happen to be in the bits which currently get priority in mathematics lessons'. In a similar vein, Dossey (1997, p. 180) observed that measurement is 'perhaps the most visible, but least considered, aspect of quantitative literacy'. Of course students encounter ideas of measurement, and of decimals, in other curriculum subjects, including geography, physical education, design, and science, yet this is all the more reason for the *mathematical* basis for measurement and decimals to be strong.

On top of this, as Kent *et al.* (2011, p. 757) conclude from their research on measurement in the workplace, students in school mathematics should not only learn 'what and how to measure attributes', but also, they argue, experience 'a careful balance with the why', the reason being that 'mathematical reasons are only one type of the many reasons taken into consideration when making decisions [in the workplace]' (also see Bakker *et al.*, 2011). This means that in school mathematics there needs to be curriculum and teaching that provides a strong mathematical basis for measurement and decimals alongside a clear focus on the use, nature, and purposes of measurement.

If measurement tends to be a neglected topic across school mathematics for older students, as seems likely, this has consequences that, according to Battista (2007, p. 902), might comprise 'the tip of a huge learning-difficulty iceberg'. Battista's contention, given the pervasiveness of measurement in geometric and graphical contexts, is that 'poor understanding of measure might be a major cause of learning problems for numerous advanced mathematical concepts' including, he suggests, graphs of functions, loci, vectors, and so on. It is for this reason, and others covered above, that this chapter focuses on connecting measurement and decimals. This is not to underestimate the importance of the more usual coupling of geometry and measurement through geometrical measures such as length, area, volume, capacity, angle, and so on, nor to ignore the importance of common fractions (see Chapter 3). Rather our aim is to show how the research evidence points to a repositioning of measurement, especially its alignment with decimal representations, as a key component of mathematics learning in the 9–19 age range that reaches out and touches all other components.

In this chapter, we first examine the nature of measurement and decimals. Then, after a review of learners' prior understandings that secondary teachers can expect to draw on, we consider possible routes of progression in measurement and decimals at the secondary school level. Following this, we summarise what is known from research about different teaching approaches that might inform classroom choices. Finally, we suggest possible issues for which more evidence is needed and where practitioners can make a particular contribution. By touching on mathematical ideas such as trigonometry, functions, calculus, dimensionality, proof, and so on, the chapter refers to notions from earlier chapters in this book and provides part of the preamble to later chapters, including the final chapter.

# The nature of measurement and decimals

Measures are numbers that are connected to a quantity. When we measure anything, we use a unit of measurement. In most practical circumstances, units of measurement are, in essence, the result of agreement. Standards of measure have evolved over time such that the measures in widespread use nowadays are mostly established through international treaties. For example, the International System of Units (abbreviated SI from *Le Système International d'Unités*) is the world's most widely-used system of measurement, both in the sciences and in everyday commerce. It is governed by the *International Bureau of Weights and Measures* to ensure worldwide uniformity and does so with the authority of the *Convention of the Metre*, a diplomatic treaty between 54 nations across the world. In the same way, precision measuring instruments are calibrated according to a globally-accepted standard, often under strict laboratory conditions, an example being the precision instruments for measuring weight that are calibrated to a globally-accepted gravity standard.

While measures are numbers, it is imperative to recognise that a measure is a ratio comparison and not a number of 'things' (Thompson and Saldanha, 2003). When whatever is being measured cannot be described solely with whole units, then there are two common strategies: one is to move to using fractions or decimals, sometimes called ordinary fractions and decimal fractions; the second is to use or devise smaller units such as inches for measures of length which cannot be described as whole numbers of feet. The use of decimals is, of course, inherent in the SI system, yet other, culturally-developed, measurement systems exist and continue to be used sometimes alongside metric measures (an example being the use of miles to measure distances between towns even when SI units such as metres are in widespread concurrent use). Even without knowing actual measurements, a relationship between two measures can be expressed quantitatively; for example, a learner might establish through comparison that one stick is three times the length of another stick, even though the individual lengths are not known. Such multiplicative comparisons are one of the roots of algebra and proportional reasoning (see, for example, Dougherty, 2008; plus Chapters 2 and 3).

Decimals are a form of place-value representation in which the position or place of a digit represents its value in base 10. Thus the 5 digit in 1500 means 5 hundred. Being taught as such, it could be thought that a decimal such as 0.75 would be easier for learners to understand than ordinary fractions. What is more, arithmetic operations with decimals might also be thought to be easier as they can be taught as further extensions of place-value representation. For instance, adding 1/2 and 3/10 is a cumbersome process whereas adding the same numbers in their decimal form (i.e. 0.5 + 0.3) would seem a simpler matter if place value representation is fully understood. The argument that decimals are easier to learn than ordinary fractions necessitates closer attention; we examine the research evidence below.

## Measurement scales

Measurement scales are commonly classified as nominal, ordinal, interval, and ratio, depending on the quantity being measured (as proposed by Stevens, 1946). Even so, there remains debate about the veracity of the classification, including whether measurement is actually entailed in uses of the nominal scale (where the use of numbers is arbitrary, as seems often to be the case with bus numbers) and the ordinal scale (where things are put in some order, like being first, second, or third in a race without knowing the precise times). In the case of the interval scale, points on the scale are considered equidistant but the zero is arbitrary.

An example of an interval scale is the measurement of temperature, illustrated in Figure 4.1. In this case, equal differences on the scale represent equal differences in how we have decided to measure temperature (whether using the Celsius or Fahrenheit scale). The absence of a 'true' zero point, however, means that it is not possible to make statements about how many times higher one temperature is than another. In other words, while the difference between 15 degrees and 20 degrees Fahrenheit is the same difference as between 35 degrees and 40 degrees Fahrenheit, a temperature of 40 degrees Fahrenheit cannot be said to be 'twice as warm' as 20 degrees. Whether 'absolute zero' on the Kelvin scale is a true zero remains an issue of debate.

**Figure 4.1** The Celsius and Fahrenheit scales of temperature

The ratio scale of measurement satisfies all the properties of measurement including equal intervals and a non-arbitrary zero. Examples of ratio scales include the measurement of length, area, volume, angle (disregarding any philosophical issues about defining zeros in each situation); the measurement of someone's age or their number of children are also examples. The notion of ratio scales almost implies that measures are always rational numbers that can be expressed as a ratio $\frac{a}{b}$, where $a$ and $b$ are integers (and $b$ is non-zero). However, measurement of the diagonal of a square, or the circumference of a circle, involves irrational numbers that cannot be expressed in that way, entailing $\sqrt{2}$ and $\pi$ respectively. As such, there are some mathematically precise measures that cannot be represented as terminating or repeating decimals. This links with the idea of approximation in measurement. Another link from irrational numbers is the distinction in measurement between discrete scales of measurement and continuous scales. Counting the number of students in different school classes is an example of a discrete scale of measurement as it makes no sense to have fractional numbers of students, either as a decimal fraction or a common fraction. In continuous scales of measure, any value can be taken. In practical situations involving use of a measurement instrument, be this a ruler for measuring length or a laser device for measuring room size, there is always the issue of approximation.

www.nuffieldfoundation.
org/measurement-2

www.nuffieldfoundation.
org/measurement-5

www.nuffieldfoundation.
org/measurement-3

## Compound measures

Compound measures, as the name implies, means that a measure relies on two other measures of different types. Examples of compound measures include measuring speed in, say, metres per second or density defined as mass divided by volume. Further categories of measure include non-linear measures and what are sometimes called derived measures. Examples of non-linear measures include the measurement of sound intensity in decibels and the intensity of earthquakes (now using the 'Moment Magnitude Scale' rather than the earlier Richter scale). Both these measures are logarithmic in that the scale of measurement uses the logarithm of a physical quantity instead of the quantity itself. For example, with the intensity of the smallest audible sound (near total silence) defined as 0 dB, a sound 10 times more powerful is 10 dB, while a sound 100 times (i.e. $10^2$) more powerful is 20 dB. An example of a derived measure is the light year, the distance that light travels in one year, almost 10 trillion kilometres, that is $10^{16}$ metres, or about 6 trillion miles. Another example of a derived measure is density as it is derived from the mass of an object divided by its volume.

In more advanced mathematics which students encounter beyond the age of 16, and certainly beyond the age of 19, there is the mathematics of infinitesimals, limits, instantaneous rates of change, accumulation, Riemann sums, and other measurement concepts that relate to differential and integral calculus. In theoretical mathematics, the study of measures and their application to integration is known as measure theory.

 www.nuffieldfoundation.org/measurement-1

 www.nuffieldfoundation.org/measurement-4

 www.nuffieldfoundation.org/measurement-3

 www.nuffieldfoundation.org/measurement-5

## Summary

Overall, given that all the above aspects of both measurement and decimals occur across the various components of the school mathematics curriculum, this means that a suitable curriculum and skilful teaching are needed if learners are to make suitable progress. Later in this chapter we review what research evidence might inform classroom choices. In the next section we outline the prior understandings that learners gain in primary school and which secondary teachers might expect to draw on and develop.

# Measurement and decimals in the primary school years

In learning measurement and decimals, there can be conceptual difficulties in understanding what is being measured as well as notational difficulties in knowing how to work with decimal notation, and these can be intertwined.

## Quantities being measured

In the early years of schooling, the development of measurement skills usually starts with directly comparing objects along one dimension. In this way, children generally succeed in length measuring prior to tackling the measurement of area and volume (Hart, 1984). Nevertheless, even length measuring poses specific problems for young learners (Battista, 2006). One source of difficulty for young learners is grasping how measurement 'is based on imagining one-to-one correspondence of iterated units' (Bryant, 2009, p. 18). This is the issue that arises when the 1 cm mark on the ruler is set at one end of the length being measured, rather than the 0 cm mark. When this happens it is likely to be because what is not clear to the learner is the difference between counting (which starts with 1) and continuous quantity which starts with zero. Research evidence (e.g. Battista, 2006; Hart, 1981, 1984) indicates that when children measure lengths they may be applying a poorly-understood procedure rather than focusing on the correspondence between the units on the ruler (which may be seen erroneously as a counting device) and the length being measured. The evidence is that children can struggle with grasping the importance of iteration (i.e. repeated units of measure) during length measurement throughout primary school, and even beyond.

Research indicates that area measurement poses further challenges for learners (see, for example, Cavanagh, 2007; Kidman, 1999; Outhred and Mitchelmore, 2000). In order to measure area, spaces are practically or hypothetically divided into equal units. Understanding this use of units of equal area turns out to be relatively difficult for learners. A particular source of difficulty is that area is routinely calculated from lengths, rather than measured using iterated units of area measure (Kamii and Kysh, 2006). In this way, area measurement turns to calculation without reference to the geometric situation and this can prove to be a further source of difficulty in that it means that the calculation procedure tends to become divorced from the concept of area measurement.

Turning to 3D measures, the measurement of volume introduces even more complexity, not only by adding a third dimension and thus presenting a significant challenge for students' spatial sense, but also in the very nature of the entity being measured (Battista, 1999, 2003, 2004, 2010; Ben-Chaim et al., 1985; Cooper and Sweller, 1989). One distinction that can be seen in elementary education is between *volume* as the amount of 3D space something occupies (perhaps measured in cubic centimetres), and *capacity* as how much liquid a container can hold (measured, say, in millilitres). As such, measuring volume might be envisaged as 'packing' a space with cubic units (Battista and Clements,

1996), while capacity is 'filling' a container with iterations of units of a fluid (or small grains such as sand or rice) that take the shape of the container (Curry and Outhred, 2005). In the latter case, research points to care being needed when classroom tasks are being designed. For example, in filling a cylindrical jar in which the linear height of the jar corresponds to the volume, the process of filling the jar with cupfuls of rice can, according to Curry and Outhred (2005), seem to learners to be simple iterative *counting* of cupfuls and not perceived by them as *measuring* the geometric 3D capacity or volume. Even so, Schmittau (2005) has shown that when comparing quantities by pouring from one vessel into another, primary school age students can learn to use algebraic ideas to express relations such as 5 cups of water are equal in quantity to 3 other-size cups, in effect that $3x = 5y$ (see Chapter 2). This illustrates that the teaching of capacity, as with volume, needs to be conducted with care and thought.

The measuring of angles, according to Bryant (2009, p. 4), is 'another serious stumbling block for pupils' (for more on this issue, see Keiser, 2004). One problem, according to Bryant, is that learners find it hard to grasp that two angles in very different contexts are the same. For example, themselves turning 90 degrees and the corner of a page in a book being 90 degrees feel like very different measurements. Similarly, students can fail to see the connection between angles in dissimilar contexts, like the steepness of a slope and how much a person has to turn at a corner. Evidence from studies such as Clements and Battista (1990) and by Clements *et al.* (1996) found that, when using a computer microworld such as *Logo*, children learnt about angle relatively well in the context of movement (though see cautions further below).

Teaching the measuring of time using analogue and digital clocks is, research suggests, generally well-established in primary school despite the complexity of the topic meaning that it can take years before students become competent (Burny *et al.*, 2009). Reading analogue clocks, in particular, is experienced by children as quite difficult compared with digital clocks. While it seems that most children are able to read both analogue and digital clocks correctly by the age of 8 to 10 (Friedman and Laycock, 1989), research by Boulton-Lewis *et al.* (1997b) indicates that it cannot be assumed that by age 11 children are necessarily competent with reading and recording the time.

## Decimal notation

In terms of decimals, the primary curriculum in some countries privileges decimal representation over fractions, while in other countries it is ordinary fractions that are given greater attention. Given that decimals are used in measurement, learning to handle decimals is necessary for primary-age children. While children in the primary years may be taught that the values of the digits in a number are structured on multiples of 10, that zero is a place-holder, and that the decimal point is a marker that separates the whole number part from the fractional part, the evidence is that these ideas take time to embed (Nunes and Bryant, 2009). As a result of the complexity of what is often called *number sense* (Greeno, 1991; Howden, 1989), including the notion of place-value, primary children can have difficulties with decimals, including difficulties in making judgements about the relative size of decimal numbers.

All this can mean that older primary students (aged 9 to 11) may have limited success when comparing decimals written with different numbers of digits after the decimal point (for example, when comparing 0.5 to 0.36). Such difficulties can relate to their understanding of what is being measured, and how the measurement is being represented in decimal notation.

## Learner progression in measurement and decimals

Analyses of mathematics curricula conducted within the TIMSS (Third International Mathematics and Science Study) (e.g. Schmidt *et al.*, 1997, p. 64) reveal that in most countries across the world, the ideas within measurement begin to be introduced in the curriculum in the early years of compulsory schooling, become a major focus when children are 8 or 9, but are not completed until students are 16 or 17 (see below for examples). The introduction of decimals to students occurs when children are 8 or 9, coinciding with the major focus on measurement. Whether this is coincidental or by design is not necessarily clear. Whatever the case, the topic of decimals is probably more or less completed by the time learners are aged 14. In this way, ideas of measuring within the typical mathematics curriculum precede ideas of decimals, and go on beyond curriculum coverage of decimals as learners meet, for example, new kinds of quantities, measuring tools, and dimensions.

Hence, in the typical mathematics curriculum, measurement might focus primarily on both the process of measuring, including the notion of being accurate and the associated skills of using measurement tools, and also on calculating various measures (length, area, and volume, for example) and on conversions between units (sometimes within the metric/decimal system of measures, sometimes between the metric system and non-decimal systems of measure). In terms of the measurement of shapes, finding the perimeter and area of circles usually comes later than working with the perimeters and areas of straight-edge polygons like squares and rectangles. Coverage of the measurement of time might involve the reading of analogue and digital clocks, and the use of a calendar, in order to determine durations, to use timetables, and so on.

For students in secondary school, further challenges come when ideas of measurement extend to a consideration of compound measures such as speed, acceleration, fuel consumption, and so on. At about this time in learner progression, there are issues to do with the imprecision of measurement (perhaps including the idea that a measurement is only 'accurate' to within a half unit in either direction), plus a need for facility with ratio and the idea of fractions as division. Later come ideas of the continuous nature of scales that are used to make measurements. Beyond the age of 16, students are likely to encounter radian measure in geometry and further statistical measures such as standard deviation and variance calculated on both ungrouped and grouped data. In terms of decimals, the typical secondary school mathematics curriculum is likely to cover decimal notation, calculating with decimals, and fraction-decimal-percent equivalences, fairly early.

 www.nuffieldfoundation.org/measurement-1

Amongst the issues to deal with in any consideration of progression in measurement and decimals are ones to do with when and how to tackle large measurements, such as distances in astronomy, and extremely small measurements, such as the mass of an electron. Further advancements are in the use of scales where zero does not mean 'nothing' (such as when measuring temperature), and measures which are linear in mathematical terms but which may not look that way (such as when measuring angle).

All told, as a report from ACME (Advisory Committee on Mathematics Education) (2011, p. 20) explains, measurement (usually represented in decimal form) requires: 'appreciation of conservation; understanding the nature and dimensionality of the quantities to be measured; understanding continuous number; understanding units and their iterative use; and knowing how to use the relevant tools'. This entails learners needing to understand how measurement relates to counting and scaling. They also need to make the shift from viewing number as discrete to appreciating number as continuous (a key idea that is known to take time and multiple experiences). On top of this there is not only the idea that ratio is embedded in measurement, since measuring compares units to other quantities, but beyond this that compound measures, such as kilometres per hour, are rates that relate to gradients (slopes) of functions.

 www.nuffieldfoundation.org/measurement-3

 www.nuffieldfoundation.org/measurement-4

It can therefore happen that in the specification of a mathematics curriculum (at, say, the national level) the idea of progression from linear measures such as length or distance, through square and cubic measures such as area and volume, to compound measures such as speed, acceleration, and so on is inherent. Research explicitly investigating the veracity of this progression of measurement ideas over such a relevant span of schooling is limited, mainly because it is difficult to extract learners entirely from this progression and investigate alternatives. In addition, specifying progression in angle may need to be handled differently because angle is neither linear in a geometrical sense, nor is it multi-dimensional or compound.

Nevertheless, Brown *et al.* (1995a, p. 168) do conclude their study of learners in the 6–13 age range by claiming that 'despite the wealth of individual variation at the detailed level, the degree of consistency in a core of general ideas [of measurement] supports the contention that a common model for progression in the form of a curriculum and assessment framework can be feasible and appropriate', though this conclusion was subject to critique (Austin, 1995; van den Berg, 1995) and response (Brown *et al.*, 1995b). The model suggested by the work of Brown *et al.* (1995a) has most 6-year-olds being able to use multiple concrete non-standard units for measurement and being able to read whole numbers to 100. At the upper age most capable 13-year-olds should be able to convert readily between metric units, and be able to use equivalences of decimals.

The detail of such a common model for progression continues to be the subject of research. For the most part, the research that has been reported to date is limited to specific student ages and largely to primary age children. For instance Gravemeijer *et al.* (2003) researched a 'hypothetical learning trajectory'

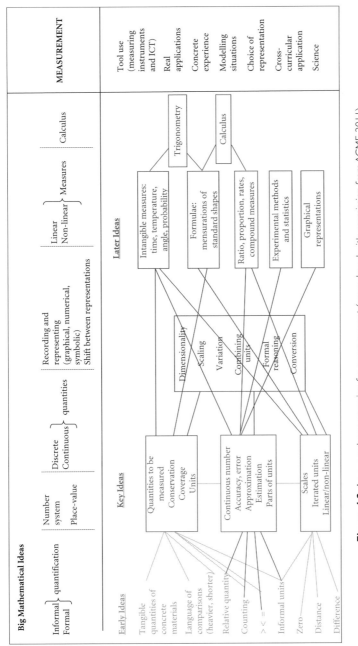

**Big Mathematical Ideas**

| Informal } quantification<br>Formal | Number<br>system<br>Place-value | Discrete } quantities<br>Continuous | Recording and<br>representing<br>(graphical, numerical,<br>symbolic)<br>Shift between representations | Linear<br>Non-linear } Measures<br>Calculus | MEASUREMENT |

**Early Ideas**

Tangible
quantities of
concrete
materials

Language of
comparisons
(heavier, shorter)

Relative quantity

Counting

> < =

Informal units

Zero

Distance

Difference

**Key Ideas**

Quantities to be
measured
Conservation
Coverage
Units

Continuous number
Accuracy, error
Approximation
Estimation
Parts of units

Scales
Iterated units
Linear/non-linear

Dimensionality

Scaling

Variation

Combining
units

Formal
reasoning

Conversion

**Later Ideas**

Intangible measures:
time, temperature,
angle, probability

Trigonometry

Formulae:
mensurations of
standard shapes

Calculus

Ratio, proportion, rates,
compound measures

Experimental methods
and statistics

Graphical
representations

**MEASUREMENT**

Tool use
(measuring
instruments
and ICT)

Real
applications

Concrete
experience

Modelling
situations

Choice of
representation

Cross-
curricular
application

Science

**Figure 4.2** An exemplar progression for measurement (reproduced with permission from ACME, 2011).

(or possible learning route or progression) that focused on measurement and what they called 'flexible arithmetic' for 6-year-olds. More recently, there have been a number of reports on possible learning trajectories for measurement for students aged between about 4 and 8; for example, Barrett *et al.* (2012), Sarama *et al.* (2011), and van den Heuvel-Panhuizen and Buys (2008).

The ACME report mentioned above (ACME, 2011, p. 21) does speculate about how it might be possible to map out for the topic of measurement 'the essential ideas, components and proficiencies, and how they link together'. The exemplar mapping that they produced for the topic of measurement is shown in Figure 4.2, illustrating progression from early ideas of informal quantification, through discrete and continuous quantities to compound and non-linear measures and beyond.

Whatever the detail of the learning progression for measurement and decimals currently enacted across countries, pupil performance data from TIMSS indicates that it is only by the age of 14 that the median pupil attainment across the world extends to understanding measurement in several settings and understanding place-value of decimal numbers (e.g. Kelly *et al.*, 2000, p. 13). What is more, it is only the top 10% of school students of that age internationally who can solve time-distance-rate problems involving conversion of measures within a system (i.e. relations between quantities and measures of rate). This illustrates that, at the beginning of their time in secondary school, learners' understanding of measurement and decimals is likely to be quite mixed and can depend very much on specific teaching foci within their primary school experience of mathematics. It is on that basis that we turn to consider evidence for teaching approaches for measurement and decimals.

## Teaching approaches for measurement and decimals

In acknowledging that the major curricula focus on measurement and decimals occurs when children are 8 or 9, it is understandable that the research focus has been with children at and around that age range. Indeed, as exemplified above in the research on learning progression, research about teaching and learning measurement has mainly been with children from the early years of schooling through to the ages of 8 or 9. Hence, in looking at teaching approaches beyond the age of 9, it is sometimes necessary to extrapolate from research conducted with slightly younger children.

An example of such research is that of Lehrer (2003) where, although his main focus is measurement with young children, he usefully lists eight components that provide a basis for the 'web of ideas' that he sees as comprising measurement across the whole of school mathematics:

- unit-attribute relations (what units can/should be used);
- iteration (subdivision into congruent parts or repetition of a unit);
- tiling (that gaps must not be left between units);
- identical units (if units are identical, then a count represents the measure);
- standardisation (using a standard unit makes communication of measures easier, and indeed possible);
- proportionality (different units can be used to measure and can be compared);
- additivity (a line segment can be divided into several smaller line segments whose sum will equal the original length);
- origin (when using a scale to measure, it is important to identify the zero point) (adapted from Lehrer, 2003, pp. 180–181).

According to Lehrer, these eight components are vital to the development of understanding of all measurement attributes, be they length, area, volume, mass, angle, time, and so on. For students, Lehrer reports, the development of a conceptual grasp of these different attributes of measures is neither simultaneous nor sequential in a straightforward way. On a more positive note, Lehrer suggests that understanding of the eight components is extendable from one component to another. For example, a student who appreciates, when measuring a length using hand spans, that it is important not to leave gaps, is more likely to understand that measuring an area using tiles must also leave no gaps. Subsequently, Lehrer *et al.* (2003) extended this set of 'central concepts' with one further idea, the addition of the concept of *precision*; all measurement is inherently approximate and that one determinant of the level of precision can be the choice of units, with the choice likely to be based on the context of the measuring being undertaken.

 www.nuffieldfoundation.org/measurement-1

 www.nuffieldfoundation.org/measurement-4

 www.nuffieldfoundation.org/measurement-3

 www.nuffieldfoundation.org/measurement-5

## Geometric measures

It seems that when teaching length measurement and ideas of unit-attribute relations and iteration of units, the problems that younger children experience (see above) appear to persist well into secondary school. For example, when Hart *et al.* (1985) showed a large number of students aged 11–14 a picture of straight line beside a ruler marked in centimetres, such that one end of the line aligned with the 1 cm mark on the ruler and the other end with the 7 cm mark (as in Figure 4.3), almost as many 11 year olds (46%) gave the answer 7 cm as gave the correct answer of 6 cm (49%).

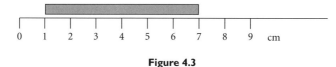

**Figure 4.3**

In another large-scale study (reported in Kloosterman *et al.*, 2004; Sowder *et al.*, 2004), learners were asked to judge the length of an object which was pictured just above a ruler with neither of its endpoints aligned with the zero endpoint on the ruler. Only about 60% of the 13-year-olds managed to find the correct answer. As these research findings illustrate, there are secondary-age school students who may know procedurally how to use a standard ruler, but they may not fully understand the nature or structure of the measurement units with which they are dealing.

With area measurement, an issue for learners is that a measure given in one kind of unit is used, often via calculation, to produce a result in another kind of unit. For example, if length is measured in centimetres then area would be in square centimetres. While such calculation of area is multiplicative, research indicates that students often attempt to calculate area by adding parts of the perimeter rather than by multiplying (e.g. Hart, 1981, 1984; Kidman, 1999). Even when students appear to have grasped the rectangular area formula (i.e. area = length × width) they can have a tendency to employ it in all contexts, regardless of the actual shape being considered (Dickson, 1989). Indeed, such a tendency can be so deeply held that it persists into adulthood (Baturo and Nason, 1996).

The large-scale study of 11–14-year-olds by Hart *et al.* (1985), mentioned above, provides evidence of the difficulties that many secondary school student continue to have with area measurement. In one task, Hart and colleagues showed 11–14-year-olds a rectangle measuring 4 squares (base) by 2.5 squares (height) drawn on squared paper and asked them to draw another rectangle of the same area with a base of 5 squares. Fewer than half of the 11-year-olds (44%) were able to provide the correct solution; a sizeable proportion judged that it was impossible to solve the problem. Students who are familiar with manipulating shapes using transformations and dissections, are likely to find this fairly simple, since doubling the 2.5 measurement and calling it 'the base' while halving the other dimension solves the problem (Figure 4.4). Hurdles to be overcome in order to do this include the belief that the words 'base' and 'height' fix the orientation of the shape and affect the area calculation.

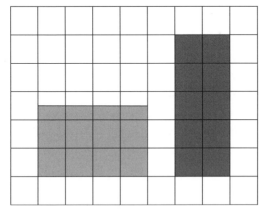

**Figure 4.4**

Another consequence of the multiplicative nature of area calculations is, of course, that doubling a figure's dimensions quadruples its area; this can be difficult for students to understand.

www.nuffieldfoundation.org/measurement-2

www.nuffieldfoundation.org/measurement-5

## The relationship between measuring and proving

The way in which area measurement formulae relate to geometric theory is considered in Chapter 5. At this point it is worth noting that a continuing issue for secondary mathematics teachers, as well as for curriculum designers, is the somewhat uneasy relationship between measuring and proving. This can be exacerbated when measurement is seen as entirely empirical, in contrast to mathematical proof which is invariably taken as solely deductive. It is clear, however, that quasi-empirical arguments can satisfy students in the 11–16 age range (Chazan, 1993; de Villiers, 1991), and probably beyond. For example, in a study reported by Patronis (1994), a class of 16-year-olds relied on angle measurements to 'prove' that the bisectors of two angles whose sum was 180 degrees were perpendicular.

Nevertheless, there are ways in which it might be possible to have a more mutually supportive relationship between measurement and proving. Chazan (1993), for example, suggests that the use of measurements in geometry teaching does not necessarily hinder students' appreciation of mathematical proof if the teacher can use measurement to show the difference between empirical and deductive reasoning. Useful in this context is research by Olivero and Robutti (2007) showing how using measurement in dynamic geometry software[1] (DGS) can be a powerful tool in supporting the teaching of conjecturing and proving, though they caution that measuring in DGS is a complex tool that requires appropriate management and interpretation. Overall, as Laborde (1998) illustrates, students tackling geometry problems move to and fro between 'spatiographic geometry' and 'theoretical geometry'. This means that when attempting a geometric proof, a student might move from making conjectures using measures taken from a geometrical drawing, to using definitions and theorems, then go back to the drawing, and so on. Other examples of 'measuring' that are closer to proving might include finding quantities by deductive reasoning about properties (in geometrical situations, this is sometime called 'angle-chasing') and, as mentioned above, the use of area formulae in a way that makes clear the relationship with geometric theory.

As a result, it is likely that many students above the age of 11 would benefit from more opportunities to build their knowledge of measures of length, mass, and capacity through, for example, solving problems which involve converting from one unit of measure to another. They also need opportunities to build understanding of the distinction between perimeter as a linear measure and area as a square measure of surface. Likewise, students are likely to benefit from more opportunities to develop their understanding of units of time and the relationship between them, and more opportunities to practise using known angle facts to recognise when a problem can be solved by reasoning rather than estimation.

---

[1] Examples include Cabri™, Cinderella™, Geogebra™, Geometer's Sketchpad™.

## The influence of tools

As with the way learners' knowledge of measures of length, area, and volume is affected by the tools employed and invented by learners, so it is with their developing conceptions of angle. Research on learners' use of the *Logo* computer microworld shows how such usage affords the idea of angle as rotation. Kieran (1986, p. 104), for example, found that 9-year-old children, after one year of using *Logo*, seemed to keep static 'angles' and their measurement in one mental compartment, and dynamic turns and their measurement in another. This means that children using *Logo* might learn about angle relatively well in the context of movement, as Clements and Battista (1990) and Clements *et al.* (1996) found, yet may not transfer this understanding to geometry tasks where they are calculating using static 'angles'.

By the time students are 11, they seem more able to integrate static and dynamic imagery of angles; Mitchelmore and White (1998, 2000) refer to this development as the abstraction of the angle concept. While there are currently a number of DGS tools available that might assist with abstraction of the angle concept, the main focus for research with such tools has often been primarily on geometric constructions and ideas of proof, rather than necessarily on the concept of angle measure. For example, although Ruthven *et al.* (2005) provide accounts of classroom tasks on angle properties, the focus for the tasks was more about early ideas of proof than about understanding angle as a measure. Even when focusing on the angle concept, a study by Lineberry and Keene (2011, p. 1031) found that while using DGS helped students (aged 15–17) to gain insight into the notion of angle as between two lines (in the static Euclidean sense), few of the students showed a better understanding of angle as rotation and none demonstrated a better understanding of angle as measuring along the arc of a curve, a precursor to radian measure.

When learning to estimate measures, a range of research suggests that students need many different experiences of estimating in order to internalise suitable benchmarks for various measures such as length, mass, angle, time, area, volume, temperature, and so on (Chang *et al.*, 2011; Jones *et al.*, 2012; Joram *et al.*, 1998, 2005). Through these different experiences, students develop strategies for making estimates and for improving the accuracy of their estimating.

 www.nuffieldfoundation.org/measurement-9

## Decimal concepts

Just as learning about measurement is a multidimensional task, then so too is learning about decimals. To be successful, students need to coordinate place-value concepts with aspects of whole number and fraction knowledge. Making the transition to understanding decimals fully relies on students having a thorough understanding of previous concepts and this becoming fully integrated with new information. As Roche and Clarke (2004, p. 486) explain, results from major research studies such as Brown (1981) and Wearne and Kouba (2000) indicate that decimals 'create great confusion for many students with much of this difficulty arising because students treat decimals as whole numbers'.

This lack of expertise with decimals persists through schooling (Steinle and Stacey, 2003, 2004). As Steinle and Stacey (2004, p. 547) say, 'Students who do not know that 0.453 is near 0.45 but a little bigger, or students who think that 0.2 is near 0.3 but a long way from 0.21345, cannot make sense of the mathematics they are being taught'.

In terms of decimal concepts, it is connecting conceptual and procedural knowledge of decimals that is vital (Hiebert and Wearne, 1986). While, as this chapter argues, this occurs through measurement, it is also important to connect decimals with percentages and with fractions (of course, there are other important contributions that knowledge of fractions makes to overall mathematical understanding, such as their role in expressing non-integer division). For example, Carpenter *et al.* (1981) advocate building a strong understanding of decimals through two approaches: capitalising on students' knowledge and skill with whole numbers, and tying their understanding of common fractions to that of decimals. To accomplish this, students need a firm understanding of the place-value system and how it relates to decimals, and a good background with common fractions to aid development of ideas of tenths and hundredths, and so on.

www.nuffieldfoundation.org/measurement-1

www.nuffieldfoundation.org/measurement-4

www.nuffieldfoundation.org/measurement-3

This approach to building understanding of decimals is supported by evidence from O'Brien (1968) who found that students who were taught decimals with an emphasis on the principles of numeration, with no mention of fractions, scored lower on tests of computation with decimals than those taught the relation between decimals and fractions. Another approach, reported by Hunter and Anthony (2003), uses percentages as a representation of decimals; this, say the researchers, provides a significant positive influence on students' developing understanding of decimals as quantities. More generally, Irwin (2001) reports on research with 11–12-year-olds that found that students who tackled tasks involving decimals in realistic contexts made much more progress in their understanding of decimals than students tackling equivalent non-contextualised tasks. As this chapter argues, it is measurement that might provide the context that helps develop students' understanding of decimals.

## Summary

This chapter illustrates the ways in which measurement is intimately tied to diverse mathematical concepts from number through to geometry, algebra, and statistics. In particular we show how ideas of measure relate to decimals. In addition, with measurement and decimals there is the obvious provision of necessary tools for scientific and other purposes outside mathematics. Without a suitable curriculum, and suitable teaching, the necessary links are unlikely to be immediately clear to learners. One of the major challenges in secondary school mathematics teaching is to find creative and imaginative ways of ensuring learners work out and benefit from such relations between mathematical ideas, and between mathematical ideas and other subjects.

Whatever the approach taken during learners' prior experience of measurement and decimals in primary school, the research reviewed in this chapter demonstrates that neither measurement nor decimals are best taught in secondary school as simple skills; rather, each is a complex combination of concepts and skills that develops over a number of years. The available evidence indicates that the principles of measurement and of decimals are not straightforward for many learners and may require more attention in school than is usually given. Measurement and decimals need a strong focus in secondary school mathematics. If this happens, the dangers mentioned at the start of this chapter (taken from Battista, 2007, p. 902) of storing up problems for learners as they move to more advanced mathematical concepts (such as functions, loci, vectors, and so on) might be tempered or avoided.

## Where additional evidence is needed

There remain many questions about the teaching and learning of measurement and decimals where evidence that can be provided by practitioners is much needed:

- This chapter argues for a stronger linking between measurement and decimals at the secondary school level. Studies could report on ways of teaching measurement and decimals that use the context of measurement to good effect in the teaching of decimals. Studies could also focus on how teachers might use students' knowledge of decimals to enhance their understanding of, say, compound measures (such as speed, acceleration, and fuel consumption).
- Digital technologies such as sensors and data-loggers for gathering and displaying measurements are increasingly available. Studies might explore the different ways such technologies can be used to help build students' understanding of measurement and decimals.
- An issue in the curriculum is ensuring a mutually supportive relationship between measurement and proving. Studies could examine how this relationship maps out in schools and how the relationship might be improved in the curriculum. A suitable focus might be on issues such as when angle problems can be solved by reasoning, or how the relationship between area formulae and geometrical theory is addressed.

# Key readings

Clements, D. H. and Bright, G. (Eds) (2003). *Learning and teaching measurement.* Reston, VA: NCTM.

This book illustrates the range of issues in the teaching and learning of measurement and provides many examples of teaching ideas and approaches.

Lehrer, R. (2003). Developing understanding of measurement. In J. Kilpatrick, W. G. Martin, and D. Schifter (Eds), *A research companion to principles and standards for school mathematics* (pp. 179–192). Reston, VA: NCTM.

This chapter provides an informed overview of research on the teaching and learning of measurement.

Owens, K. and Outhred, L. (2006). The complexity of learning geometry and measurement. In A. Gutiérrez and P. Boero (Eds), *Handbook of research on the psychology of mathematics education: Past, present and future* (pp. 83–115). Rotterdam: Sense Publishers.

This chapter provides a useful summary of research on the geometric aspects of measurement.

CHAPTER 5

# Spatial and geometrical reasoning

## Introduction

Geometry is one of the longest established branches of mathematics and remains one of the most important. If anything, geometry is becoming more important across many fields. This is not only because of the wide-ranging applications of geometry in everything from robotics to CGI (computer generated imagery) movies, from crystallography to architecture, from neuroscience to the very nature of our universe, but also because new geometrical ideas are being generated within such diverse fields as these (Malkevitch, 2009; Whiteley, 1999). Indeed, it could be that as much new geometry is being devised by geometers working outside of established university mathematics departments as within.

With its origins in the surveying of land and in the design of religious and cultural artefacts, the development of geometry can be traced back through a wide range of cultures and civilisations. Over the past 200 years, the nature of geometry has expanded enormously; so much so that nowadays it is possible to classify more than 50 geometries (Malkevitch, 1992). While this illustrates the richness of modern geometry, at the same time it creates a fundamental problem for curriculum designers in terms of what geometry to select to include in the school mathematics curriculum, and hence what to exclude (Jones, 2000a). In addition, geometrical richness provides significant challenges for teachers because it means that teaching geometry well involves knowing how to recognise interesting geometrical problems and theorems, appreciating the history and cultural context of geometry, and understanding the many and varied uses to which geometry is put and through which it is generated (Jones, 2002). On top of this, efforts to teach geometry well may not always be well-supported by textbooks, especially in countries where textbook quality is variable and attention is primarily on declarative aspects such as names and properties of shapes and, later, on a specific format for laying out geometric proof (Brantlinger, 2011; Herbst, 2002; Oner, 2008)

Various solutions have been put forward to solving the problem of what geometry to include in the school mathematics curriculum (Galuzzi *et al.*, 1998; Price, 2003; Sinclair, 2008; Usiskin, 1987). What is noteworthy in these efforts at reforming school geometry is that they are longstanding and that no single solution solves all the problems. In the early 1970s one well-known curriculum team even commented 'Of all the decisions one must make in a curriculum development project with respect to choice of content, usually the most controversial and the least defensible is the decision about geometry' (The Chicago School Mathematics Project staff, 1971, p. 281). Indeed, the oldest association of teachers in the world began in 1871 in the UK as the *Association for the Improvement of Geometrical Teaching*, later to become *The Mathematical Association*. That illustrates just how long there have been efforts to improve geometry teaching, and provides an indication of the scale and depth of the issues.

In more recent times, the development of digital technologies for geometry learning from microworlds such as Logo to various dynamic geometry environments (DGS) such as Cabri™, Cinderella™, Geogebra™, and Geometer's Sketchpad™, have introduced new classroom potentials. These bring into sharp relief some fundamental questions about teaching methods and the sequencing of curriculum topics (for overviews see, for example, Hollebrands *et al.*, 2008; Laborde *et al.*, 2006). The potential of digital technology for geometry education is a thread that permeates this chapter.

What is clear from the research evidence, and hence what this chapter argues, is that geometry education at the secondary school level needs to attend to two closely entwined aspects of geometry across both 2D (plane) and 3D (solid) geometry: the spatial aspects, and the aspects that relate to reasoning with geometrical theory. The former involves spatial thinking and visualisation, while the latter involves deductive reasoning using approaches that employ, as appropriate, transformation and/or congruency arguments. These twin aspects of geometry (the spatial and the deductive) are not separate; they are interlocked, they are the *yin yang* of geometry education. It is this entwining that is behind why mathematical diagrams can help learners understand mathematical ideas, proofs, and arguments; indeed, this is why the spatial and the deductive can be combined as 'proofs without words' and as 'visual proofs' (Alsina and Nelsen, 2006; Davis, 1993; Nelsen, 1993). For example, Figure 5.1 provides a proof of the Pythagorean theorem (for another such proof, one attributed to the *Chou pei suam ching*, a Chinese document dating from around 200 BCE that uses only translations of triangles within a square, see Alsina and Nelsen, 2006, p. 27).

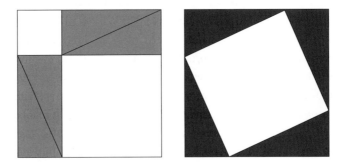

**Figure 5.1** A proof of the Pythagorean theorem.

This twinned power of spatial and geometrical reasoning is captured in the words of the mathematician Sir Michael Atiyah:

… spatial intuition or spatial perception is an enormously powerful tool and that is why geometry is actually such a powerful part of mathematics - not only for things that are obviously geometrical, but even for things that are not. We try to put them into geometrical form because that enables us to use our intuition. Our intuition is our most powerful tool… (Atiyah, 2001, p. 658).

In this chapter, we first examine the nature of spatial and geometrical reasoning. Then, after a review of students' prior understandings that secondary teachers can expect to draw on, we consider possible routes of progression in spatial and geometrical reasoning at the secondary school level. Following this, we summarise what is known from research about different teaching approaches that might inform classroom choices. Finally, we suggest possible issues for which more evidence is needed. By picking up on mathematical ideas of measurement, ratio, proportion, and algebra, and by pointing to topics such as trigonometry, calculus, dimensionality, and proof, the chapter refers to notions from earlier chapters in this book and provides part of the preamble to the final chapter.

# The nature of spatial and geometrical reasoning

A useful contemporary definition of geometry is one attributed to the British mathematician, Sir Christopher Zeeman: 'geometry comprises those branches of mathematics that exploit visual intuition (the most dominant of our senses) to remember theorems, understand proof, inspire conjecture, perceive reality, and give global insight' (Royal Society, 2001, p. 12). This definition encapsulates what can be thought of as the dual nature of geometry in that it is both one of the most practical and reality-related components of mathematics *and* it is an important area of mathematical theory. This means, on the one hand, that geometry can be seen all around us (and is widely utilised in art, design, architecture, engineering, and so on); on the other hand, it is a theoretical field that allows geometers and other mathematicians, together with cosmologists and other scientists, to work with hypothetical objects in *n*-dimensional space using, amongst other things, mathematical visualisation techniques with high-powered computers (Banchoff, 1990; Roseman, 1998).

The notion of 'figural concept' (Fischbein, 1993; Fischbein and Nachlieli, 1998) captures the combined role of the figural and the conceptual in geometry. This means that in 'seeing' a circle represented on paper, or on a computer screen, what we see is a textual representation of something which is an element of geometrical theory. One way to work with this dual nature of geometry is to distinguish between a 'drawing' and a 'figure' (Parzysz, 1988) in that, as Laborde (1993, p. 49) explains, 'drawing refers to the material entity, while figure refers to a theoretical object'. Another way is to follow Phillips *et al.* (2010, pp. 3–4) and take a geometric diagram as 'an unusual thing in that it is not an abstraction of an experienced object. Rather, it is an attempt to take an abstract concept and make it concrete'. In this sense, the term 'geometric diagram' is being used by Phillips (and by Laborde, 2004) to capture the idea that a geometric object that we see is both a material 'drawing' and a theoretical 'figure'. For example, as Laborde (2004, p. 160) illustrates, the parallelism of opposite sides of the parallelogram in the diagram in Figure 5.2 relates to the figure aspect of the diagram; the theoretical aspect. Entwined with this theoretical aspect is the material aspect of the drawing; perhaps that two of its sides are horizontal (given a conventional orientation of this book).

**Figure 5.2** A diagram of a parallelogram in a conventional orientation.

Ever since the third century BCE, or thereabouts, when much of the then accumulated knowledge of geometry was codified in a text that became known as Euclid's *Elements*, geometrical reasoning has been synonymous with the deductive method. It is through this method that, beginning with a strictly limited number of axioms and postulates, a structure of theorems is proved by deductive logic. Of course, ideas of proof, and the closely linked ideas of definitions, permeate all mathematics. Even so, in school mathematics it remains the case that geometrical reasoning is often taken to mean primarily, or even solely, deductive proving. Yet doing so ignores the process by which new mathematics is created through posing and solving problems, analysing examples, making and revising conjectures, searching for classes of counter-examples, and so on (Lakatos, 1976).

Spatial reasoning is defined by Clements and Battista (1992, p. 420) as 'the set of cognitive processes by which mental representations for spatial objects, relationships, and transformations are constructed and manipulated'. As such, spatial reasoning is a form of mental activity which makes possible the creation of spatial images and enables them to be manipulated in the course of solving practical and theoretical problems in mathematics. This links to visualisation, something which is generally taken as 'the ability to represent, transform, generate, communicate, document, and reflect on visual information' (Hershkowitz, 1989, p. 75). Both spatial reasoning and visualisation play vital roles not only in geometry itself and in geometry education, but also more widely in mathematics and in mathematics education (Giaquinto, 2007; Jones, 2001; Newcombe *et al.*, 2012; Presmeg, 1986).

 www.nuffieldfoundation.org/geometry-1

 www.nuffieldfoundation.org/geometry-2

# Theories of spatial and geometrical reasoning

In addition to Fischbein's 'figural concept' noted above, influential research-ers on the nature of spatial and geometrical reasoning, and its development in learners, include (in chronological order) Piaget (e.g. Piaget and Inhelder, 1948/1956; Piaget *et al.*, 1948/1960), van Hiele (1959/1984, 1986), and Duval (1995, 1998), amongst others. Here we only have space to give a brief outline of each; much more detail can be found in Battista (2007, pp. 846–65), Clements and Battista (1992, pp. 422–432), and Owens and Outhred (2006, pp. 84–90).

Perhaps the most appropriate summary of the research of Piaget and col-leagues is that provide by Piaget himself (Piaget, 1953, p. 75):

A child's order of development in geometry seems to reverse the order of historical dis-covery. Scientific geometry began with the Euclidean system concerned with figures, angles and so on; developed in the 17th century the so-called projective geometry (dealing with problems of perspective); and finally came in the 19th century to topology (describing spatial relationships in general qualitative way...). A child begins with the last; his (*sic*) first geometrical discoveries are topological.... Not until a considerable time after he has mas-tered topological relationships does he begin to develop his notions of Euclidean and pro-jective geometry. Then he builds those simultaneously.

Hence the work of Piaget and colleagues suggests that an infant's early spatial conceptions are topological in nature, an example being when able to copy a drawing of one circle inside another, but reproducing a triangle as a cross of two separate lines. Beyond the age of 4, according to Piagetian research, children begin to perceive and represent objects from different points of view, thereby incorporating ideas of perspective. It is after the age of 9, suggests Piagetian research, that children gradually acquire the concepts of angle and parallelism that are central to Euclidean geometry.

Subsequent research has delved into the veracity of this order of develop-ment. Overall, according to Clements and Battista (1992, pp. 425–426), 'while not totally disproven, the topological primacy theory is not supported' by subse-quent research. As such, it seems that it is not altogether appropriate to arrange a school geometry curriculum in which students construct first topological and later projective and Euclidean ideas. Rather, the challenge continues to be how to devise a curriculum so that geometric ideas of all types develop over time in a coherent way.

The van Hiele model of geometrical thinking suggests that learners advance through perhaps as many as five levels of thought in geometry. These levels were characterised by van Hiele (1959/1984) as levels 0 to 4. More recently, research has begun numbering the levels 1 to 5; for example, Clements and Battista (1992, pp. 427) argue for this 'for consistency's sake'. The following characterisation of the levels uses the more contemporary 1 to 5 numbering and is informed by the original articulation in van Hiele (1959/1984).

- Level 1: appearance. At this initial level, figures are judged by their appearance.
- Level 2: properties. At level 2, figures are a collection of properties.
- Level 3: ordered properties. At level 3, properties can be deduced one from another but the intrinsic meaning of deduction is not understood.
- Level 4: deduction. At this level, ideas of deduction such as the converse of a theorem, and ideas of necessary and sufficient conditions, are understood.
- Level 5: axiomatics. At the fifth level, reasoning is about formal mathematical systems such as the establishment, elaboration, and comparison of axiomatic systems of geometry.

Alongside these five levels of thought in geometry, the van Hiele model also includes a five-phase approach to teaching geometry. The approach to teaching is based around the idea that progression from one van Hiele level to the next is more dependent upon the teaching that the learner receives, and the particular concepts being taught, than on the learner's age. Research on the model is generally supportive of the levels of geometrical thought being a useful way of describing students' geometric concept development (e.g. Burger and Shaughnessy, 1986; Cannizzaro and Menghini, 2006; Fuys *et al.*, 1988), though there is evidence that the levels may not be as discrete as the model envisages (Gutiérrez *et al.*, 1991; Lehrer *et al.*, 1998). There is a lack of research on the five-phase approach to teaching that is integral to the model.

Duval (1995, 1998) approaches geometry from a perceptual and cognitive viewpoint. In his framework he identifies three kinds of cognitive process: *visualisation processes* (for example, the visual representation of a geometrical statement), *construction processes* (when using tools, be these ruler and compass, or computer-based), and *reasoning processes*. While these three cognitive processes can be performed separately, Duval (1998, p. 38) argues that they are 'closely connected and their synergy is cognitively necessary for proficiency in geometry'. A useful aspect of Duval's work is how it relates to the use of diagrams in geometry teaching and learning. As Gal and Linchevski (2010, p. 180) illustrate using Duval's ideas, students' difficulties can occur as a result of their 'spontaneous processes of visual perception' which can, at times, 'contradict the geometric concepts/knowledge aimed at by the teacher and the tasks'. More research would be helpful on ways of enabling the synergy to occur between the three kinds of cognitive process that Duval argues are cognitively necessary for proficiency in geometry.

## Summary

In summary of these theoretical ideas, while the details vary and continue to be the subject of research, what these researchers suggest about the nature of spatial and geometrical reasoning is that various types of geometric ideas, both spatial and theoretical, appear to develop over time, becoming increasingly integrated and synthesised. At school level, relevant geometrical ideas include symmetry, invariance, transformation, similarity and congruence. All these geometrical ideas relate to the more global mathematical ideas of proof and proving, as well as to definitions and defining. The teaching of such issues is considered in more detail below.

Importantly, a growing strand of research has been examining the influence of the use of various classroom artefacts on the development of spatial and geometrical reasoning. Of particular interest is research on the impact of computer-based tools. Such research, some of which is summarised below, tends to show that when such software is integrated intelligently with curriculum and pedagogy, then measurable learning gains are produced. Even so, it can be difficult to tease out whether such gains are the direct result of using the specific digital technology or are the consequence of the re-thinking of curricula and pedagogy (Jones, 2005, 2012). This conundrum is, of course, not restricted to geometry education but occurs across topics in mathematics teaching because the tools being used, and the changed nature of teacher–student interactions, alter the nature of the questions that can be tackled and the prior knowledge that is necessary.

# Spatial and geometrical reasoning in the primary school years

Learning geometry during the pre-school and primary school years involves classroom activities that engage learners in visualising, drawing, making, communicating, and so on, about 2D and 3D shapes (Levenson *et al.*, 2011; Roth, 2011). During these years, much geometry teaching focuses on the development of language for shape (for example, the names of polygons) and for location (for instance, left and right). Of course, knowledge of mathematical terminology is essential for modelling, visualising and communicating in all areas of mathematics. Even so, the problem can be that a heavy emphasis on descriptive language and definitions, even if relatively informal, at the expense of geometrical problem solving, might mean that progression in geometry during their primary school years is somewhat limited (Clements, 2003, pp. 151–152; Jones and Mooney, 2003). This can be exacerbated when the curriculum is dominated by stipulations around the 'four rules' of number, as if competency with these is the sole criterion for successful transition to secondary mathematics. One argument in this chapter is that geometrical ideas are just as crucial to learners' continuing with mathematics.

The TIMSS study (for example, Mullis *et al.*, 2008, p. 207) reveals that, during their primary school years, most young learners across the world learn something about some properties of geometric shapes such as simple polygons and polyhedra. They also learn about lines in the 2D plane, such as being parallel or perpendicular, and about the sizes of angles and how to draw them. In contrast, it seems that learning about relationships between 3D and 2D shapes is much less common. Within geometric measurement, the calculating of the perimeters and areas of squares and rectangles is common, as is finding areas by covering with shapes or counting squares. Estimating areas and volumes is much less common, with topics within the mathematics of location and movement using, for example, informal coordinate systems the least common. Line symmetry, and transformations such as reflections and rotations, essential for many design and engineering applications, are also less frequently taught, except as generators of patterns, than the basic properties and names of lines and shapes. In general, it seems at this level there is limited coverage of geometrical ideas in most countries around the world.

What is not always taken fully into account is that children do come to school with a good deal of knowledge about spatial relations, primarily because we inhabit a spatial world surrounded by spatial objects. This means, as Bryant (2009, p. 3) puts it, that 'one of the most important challenges in mathematical education is how best to harness this implicit knowledge in lessons'. For some examples of how this can be achieved at the primary school level, see Lehrer *et al.* (1999).

Through primary school, while young children may learn the names of simple shapes (though see below for some cautions regarding the influence of prototypical representations), it can be more difficult for them to recognise the relation between transformed shapes through rotation, reflection, and enlargement. For example, primary school children are likely to need a lot of experience with transforming shapes before they could complete the rotation or reflection patterns in Figure 5.3. Perhaps this is because children's earlier experiences of mathematical shapes can focus on enabling them to recognise the same shape whatever its location or size (e.g. this shape is a square no matter what size it is) rather than also helping them to be aware of relevant transformations of the shape (Bryant, 2009). Research also indicates, as shown in the evidence presented in Chapter 4, that children experience particular problems with measuring lengths and areas, even though they may understand the underlying logic of measurement. Similarly, learning how to represent angle mathematically is not straightforward for young children, even though angles occur everywhere in their everyday life.

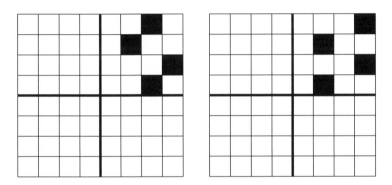

**Figure 5.3** These patterns could be completed by rotation or reflection.

When young children are learning about 2D and 3D shapes, research has documented the ways in which they are likely to do some or all of the following: under-generalise by including irrelevant characteristics that inhibit generalisation, over-generalise by omitting key properties with a result that their generalisation is too wide, and incur language-related misconceptions (e.g. that 'diagonal' means 'slanting'). In a summary of such research, Hershkowitz (1990, p. 82) shows how, for learners, each geometric object has 'one or more prototypical examples that are attained first' that are 'usually the subset of examples that had the "longest" list of attributes of all the critical attributes of the concept and those specific (non-critical) attributes that had strong visual characteristics'. For example, learners are much better at recognising isosceles triangles that are 'standing on their base' compared to those presented in a different orientation. As such, they may have difficulty deciding which, if any, of the triangles in Figure 5.4 are isosceles.

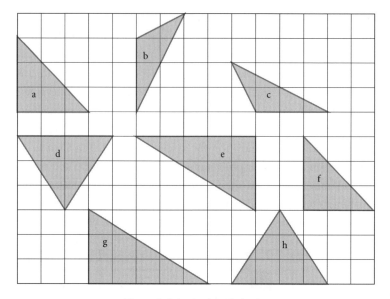

**Figure 5.4** A selection of triangles.

Other issues that learners encounter related to naming shapes (and lines) are linked to matters of definition, and to learners' embryonic understanding of necessary and sufficient conditions, and of inclusivity in defining (see below for more on issues of definition and defining). Examples of such issues include use of terms such as 'oblong' (for a rectangle that is not a square) and 'diamond' (for a specific orientation of a rhombus that is almost certainly a square), and the confusion between 'regular' and 'symmetrical'.

# Learners' progression in spatial and geometrical reasoning

In school mathematics curricula across the world, it is common for the geometry component of secondary school mathematics to focus mostly on the geometry of the plane (sometime called the Euclidean plane) and deal with objects such as points, lines, triangles and other polygons, and circles (Hoyles *et al.*, 2002). There is also likely to be some solid geometry, dealing with 3D objects such as various polyhedra (cube, tetrahedron, etc.) and perhaps the sphere. Threaded through this plane and solid geometry is attention to proof, more often solely within plane geometry. More unusual would be consideration of 'non-Euclidean' geometries such as spherical geometry, dealing with objects like the spherical triangle and spherical polygon, or, even more unlikely, hyperbolic geometry. Other significant branches of geometry, such as projective geometry or topology, are unlikely to get any mention, although attempts to include elementary topology have taken place.

In dealing with 2D and 3D shapes, the geometry component of the school mathematics curriculum generally includes the measurement and calculation of geometrical entities such as length, area and volume, and also of angle. Issues of measurement, especially as they relate to decimal notation, are considered in an earlier chapter of this book. Linking with measurement and ratios of measures, some of the consideration given to triangles and circles in the typical geometry curriculum leads to trigonometry and to calculus. In this way, the geometry component of the curriculum tends towards analytical geometry as learners progress, perhaps with graphs of functions being treated as geometric objects that can be transformed geometrically (see Chapter 8). At around this stage of education, the beginnings of vector geometry are introduced.

None of these apparent commonalities across geometry curricula should be taken to indicate that there is any uniformity in curricula progression. Indeed, Hoyles *et al.* (2002, p. 130) found considerable variation across the eight countries with which they compared the geometry curriculum of England. Amongst the variations they found was the extent to which the geometry was primarily 'practical' (exemplified at the time by the Netherlands) or distinctly 'theoretical' (for example, France or Japan), with the latter approach including more emphasis on deductive proofs. Even where the approach was 'theoretical', there nevertheless existed variations, with some countries using congruence as a central element while others used geometrical transformations.

Needless to say, and as Usiskin (1987) has argued, it is neither sensible to exclude Euclidean geometry from the school curriculum (and replace it with linear algebra, as was suggested in the 1960s), nor teach formal Euclidean geometry solely as a theoretical construct. Rather, as the evidence in this chapter illustrates, curriculum specifications need to keep in mind that spatial and geometrical reasoning develop together, and that ideas both of congruency and transformations are important during the secondary school years and beyond. The way to ensure that spatial and geometrical reasoning develop together is to plan for student progression in terms of the key *geometrical* ideas of symmetry, invariance, transformation, similarity, and congruence, and use these ideas to enable learners to build sound spatial and geometrical reasoning skills. This entails ensuring a suitable fit between the spatial aspects and those that relate to reasoning with geometrical theory in a way that embraces issues of definitions and defining, and proof and proving.

www.nuffieldfoundation. org/geometry-1

 www.nuffieldfoundation.org/ geometry-4

## Proof and proving

What is important is that a spatial and perceptual focus is balanced by the development of rigorous argument. One issue to be confronted in the mathematics curriculum in general, and in the geometry component in particular, is that it is clear that empirical arguments can sometimes satisfy students in the 11–16 age range (de Villiers, 1991; Harel and Sowder, 2007; McCrone and Martin, 2009). For example, Patronis (1994) reported that a class of 16-year-olds relied on angle measurements to 'prove' that the bisectors of two angles (whose sum was 180 degrees) were perpendicular. Research on dynamic geometry software is continuing to delve into whether the opportunities to 'see' mathematical properties onscreen might reduce (or even replace) any motivation for proof, or, on the contrary, whether DGS might open up new meaningful approaches to learning to prove (Hoyles and Jones, 1998).

As is clear from what has already been said in this chapter, during their secondary school years it seems that students continuously move back and forth between what Laborde (1998) calls 'spatio-graphic geometry' and 'theoretical geometry'. This means that when attempting a geometric proof, a secondary school student might move from making conjectures using measures taken from a geometrical drawing, to using definitions and theorems, then go back to the drawing, and so on. This moving between 'spatio-graphic geometry' and 'theoretical geometry' relates to the issue of the sometimes uneasy relationship between measuring and proving in geometry. This uneasy relationship exists even though in area measurement, for example, precise solutions can be obtained by considering the theoretical relationships between geometric shapes. For example, the 'base × height' rule for the area of rectangles applies in the same way to parallelograms and this can be proved by transforming a rectangle into a parallelogram with the same height and base knowing that the transformation does not change the area.

Similarly, the rule for finding the area of a triangle, Area = ½ (base × height), can be justified by the fact that every triangle can be transformed into a parallelogram with the same base and height by doubling that triangle. Thus, rules for precise area measurement via formulae are built from the theoretical relations between geometric shapes. While the relationship between measuring and proving is considered further in Chapter 4, it is worth noting, as Bryant (2009, p. 22) confirms, that more research is needed on learners' understanding of this centrally-important aspect of geometry and measurement.

 www.nuffieldfoundation.
org/geometry-7

 www.nuffieldfoundation.org/
geometry-8

## Geometric definitions

At secondary school level, and even beyond to undergraduate level, learners can experience difficulties in using definitions appropriately and may not fully appreciate the role of definitions in geometry (Edwards and Ward, 2004; Usiskin and Griffin, 2008; Vinner, 1991). Yet, as Freudenthal (1971, p. 424) pointed out 'Though the teacher can impose definitions...this means degrading mathematics to something like spelling, ruled by arbitrary prescriptions' (see, also, Freudenthal, 1973, chapter XVI). As such, one way to overcome such issues is for students to be actively engaged in the defining of geometric objects, as exemplified by de Villiers (1998) in the case of quadrilaterals.

In terms of similarity, Friedlander and Lappan (1987, p. 36) list a range of mathematics that is related, including enlargement, scale factor, projection, area growth, and indirect measurement. These, say Friedlander and Lappan, are 'frequently encountered by children in their immediate environment and in their studies of natural and social sciences' (1987). What is more, similar geometric shapes provide helpful mental images of ratios and equivalent fractions, and provide a model for some rational number concepts. Ideas of similarity extend to trigonometry and to the notion of self-similarity that is characteristic of fractal geometry (Senk and Hirshhorn, 1990).

 www.nuffieldfoundation.org/geometry-5

## 3D geometry

Amongst the unresolved issues in sequencing topics in the geometry curriculum is progression in 3D geometry and how this links with progression in 2D geometry. While young children might learn to recognise 3D objects, later progression might be in terms of 2D representations of 3D shapes, possibly with some 3D coordinate geometry and some 3D applications of the Pythagoras Theorem. Even as digital technologies are beginning to be used to teach 3D geometry (examples include Accascina and Rogora, 2006; Christou *et al.*, 2006), the conundrum is being raised of interpreting 3D geometry on a flat (2D) computer screen. Not only might working with 3D objects on a 2D screen seem a rather odd thing to do (compared to handling actual 3D objects), but it also raises the issue of how 3D objects are represented on a 2D screen and how the images are interpreted by the viewer (Jones *et al.*, 2009). Such perceptual issues regarding 3D objects are further complicated by the fact that the retinas of our eyes are, in essence, 2D. As a consequence, as Pizlo (2008) explains in detail, interpreting actual 3D objects is complicated enough in itself; that the 3D object appears on a 2D computer screen adds a further issue of interpretation. The development of 3D displays, either screen or projected, opens up new but as yet unexplored possibilities for geometry education.

 www.nuffieldfoundation.org/geometry-2

 www.nuffieldfoundation.org/geometry-3

## Teaching approaches to spatial and geometrical reasoning

In teaching geometry, a working group of the Royal Society (2001) has suggested that the aims include: developing spatial awareness, geometric intuition, and the ability to visualise; developing knowledge and understanding of geometrical properties and theorems; encouraging conjecturing, deductive reasoning and proof; developing skills of applying geometry through modelling and problem solving in real world contexts; and an awareness of the historical and cultural heritage of geometry in society, and of the contemporary applications of geometry. Here again is the blending of spatial and the geometrical reasoning. For the latter, the NCTM (2009, pp. 55–56) provide a useful summary of the 'key elements of reasoning and sense making with geometry'; these are:

*Conjecturing about geometric objects.* Analyzing configurations and reasoning inductively about relationships to formulate conjectures.

*Construction and evaluation of geometric arguments.* Developing and evaluating deductive arguments (both formal and informal) about figures and their properties that help make sense of geometric situations.

*Multiple geometric approaches.* Analyzing mathematical situations by using transformations, synthetic approaches, and co-ordinate systems.

*Geometric connections and modeling.* Using geometric ideas, including spatial visualization, in other areas of mathematics, other disciplines, and in real-world situations.

## Key geometric ideas

Bearing in mind the geometrical ideas of symmetry, invariance, transformation, similarity, and congruence, there are many reasons, as Freudenthal (1971, p. 434) explains, why a focus on symmetries is a good idea. He argues, for example, that the idea that something has happened to a figure if it has only been moved to another place is not necessarily clear to learners of geometry. If a cube, for example, is translated, nothing seems to have happened; if it is turned and put on a vertex, then it is more obvious that something has changed. Hence Freudenthal's view is that mirror reflection provides the strongest feeling that something has occurred. Thus teaching approaches that place symmetry as a central idea provide a basis for learner progression. This is the case not only for geometry where students meet point symmetry and line symmetry, but also for mathematics more widely. Across algebra, trigonometry, and calculus, for example, as Dreyfus and Eisenberg (1990, p. 53) illustrate, students can be 'introduced to generalisations on the notions of symmetry'. These include, symmetric functions and forms in algebra, such as the symmetric form of the equation of a line, symmetric relations in trigonometry such as $\cos A = \sin(90 - A)$, and how in calculus a pivotal role is played by symmetry when applying integration techniques and when undertaking differentiation.

In arguing for the importance of ideas of symmetry, this is not to say that symmetry is simple or uncomplicated to teach. Research tracing the development of students' knowledge of symmetry in school geometry, such as that by Leikin *et al.* (2000), has revealed a range of difficulties that learners encounter with ideas of symmetry. These range from straightforward errors, such as identifying an incorrect symmetry axis or failing to recognise a correct symmetry axis, to difficulties with reflecting in oblique lines. There are matching difficulties when secondary school students work with symmetry in three dimensions (Cooper, 1992). Research with suitable digital technologies is providing examples of how students might gain a more multifaceted appreciation of symmetry (e.g. Hoyles and Healy, 1997; Clements *et al.*, 2001).

 www.nuffieldfoundation.
org/geometry-1

 www.nuffieldfoundation.org/
geometry-4

Invariance, like symmetry, while a central idea of mathematics in general, is especially relevant and important in geometry. Hence it is especially relevant and important in the teaching of geometry. Most theorems in geometry can be seen as resulting from the study of what change is permitted that leaves some relationships or properties invariant. For example, the angle in a semicircle, as shown in Figure 5.5, is an invariant 90 degrees provided that the point being moved remains on the circle; if it moves inside the circle, the angle is greater than 90 degrees, if it is outside it is less, which could mean that a valid definition of a circle would be the locus of points that subtend an angle of 90 degrees to the ends of a line segment.

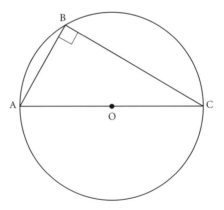

**Figure 5.5** The angle in a semi-circle.

As Schuster (1971, p. 82) explains 'invariance is one of the most important ideas in all of mathematics, and geometry is unquestionably the most natural subject for the demonstration and use of this idea'. In teaching, the use of transformations can be a means by which ideas of invariance can be studied most easily and by which the formal definitions of similarity and congruence can be related to learners' previous intuitive ideas. Here again is a place where digital technologies such as DGS can play a valuable role (see Hollebrands *et al.*, 2008; Laborde *et al.*, 2006).

## Transformations

Transformations were first introduced in a serious way into the school mathematics curricula during the reforms of the 1960s. This timing was probably unfortunate because it was at precisely the same time as the overall amount of geometry was being reduced and when much more of school mathematics was being devoted to the ideas of functions. The consequence of this timing was that the study of transformations was, at that time, less related to learners' previous geometrical experience and intuitive knowledge and more related to transformations as mathematical operations. In some ways this has fuelled an apparent ongoing dichotomy between an approach to geometry that is based on congruent triangles (in the Euclidean tradition) and the use of transformation-based arguments. Yet, as Johnston-Wilder and Mason (2005, p. 155) explain:

One of the breakthroughs in modern mathematics came as a result of switching attention from symmetries of objects to invariants of transformations of those objects. It turns out to be even more fruitful to study transformations between objects, and even to study the effects of transformations on the whole space, not just on individual objects. For example, the relation 'is congruent to' is an assertion of the existence of a particular kind of transformation (an *isometry* or distance-preserving transformation) which takes any one shape to another; the relation 'is similar to' (in the mathematical sense) is an assertion of a transformation which preserves shape but not necessarily size. These relationships lie hidden at the heart of Euclidean geometry, but it is only in the last 200 years that people have thought explicitly in terms of transformations.

Mathematically, as Barbeau (1988), Nissen (2000), and Willson (1977) demonstrate, there is nothing to choose between methods based on congruent triangles and those based on transformations. Taking an isometry as a transformation that preserves congruence, any proof by congruence can be translated into a proof by transformations, and vice versa. One written version may be neater or shorter than the other, but, in practice, neither approach is the sole purveyor of elegant proofs. Experience of more than one approach is, according to Meserve (1967), 'a necessary step in the obtaining of sufficient understanding to apply geometrical concepts effectively to mathematical problems'. This echoes with the more recent view of the NCTM (2009, p. 55) that a key element of reasoning and sense making with geometry includes using multiple geometric approaches through 'analyzing mathematical situations by using transformations, synthetic approaches, and co-ordinate systems'.

www.nuffieldfoundation.org/geometry-4

## Geometric diagrams

Given the central idea of the 'figural concept' in geometry, it is important to acknowledge the influence on learners of what are generally called prototypical geometric diagrams, such as an isosceles triangle seeming to need to stand on its short side (see Hershkowitz, 1990, p. 82). In relation to this, Askew and Wiliam (1995, pp. 14–15) commend the use of 'only just' examples and 'very nearly' non-examples of geometrical shapes, with an 'only just' example being one for which any change causes it to become a non-example, and a 'very nearly' non-example needing but one further adjustment in order to become an example. They suggest that coupling 'only just' examples with 'very nearly' non-examples is more effective than solely providing examples. Asking learners to construct such boundary examples by adding or removing constraints is also recommended by Prestage and Perks (2001) who use the construction of rhombi on coordinate grids as their example, and by Watson and Mason (1998).

www.nuffieldfoundation.org/geometry-6

## Proof and proving

A report on geometry education by the Royal Society (2001, p. 9) argues for classroom approaches that incorporate the use of logical arguments that build on what is already known by learners in order to demonstrate the truth of some geometrical result, preferably based around something previously conjectured by learners after conducting a well-chosen experiment. The report suggests that geometrical situations (i.e. the theorems) 'should be chosen, as far as possible, to be useful, interesting and/or surprising to learners'. The report goes on to suggest that the 'level of sophistication expected in the logical argument will depend upon the age and attainment of the learners concerned, and the proof produced might equally be called an "explanation" or "justification" or "reason" for the result'. To achieve this, the report suggests that learners should be encouraged to be 'critical of their own, and their peers', explanations as this should help them to develop the sophistication and rigour of their arguments'.

Such an approach, the report suggests, should mean that all learners come to understand that deductive reasoning is more than simply stating a belief or checking that the result is valid in a number of specific cases. However, 'it is not an easy matter to determine how to achieve this with each pupil and each result and that a careful choice of approach will be needed' (Royal Society, 2001, p. 9).

Of course, the process of deductive reasoning and proving must begin somewhere. The starting point for abstract mathematics is a minimal collection of initial reasonable assumptions, called axioms. In the context of school mathematics, as the Royal Society report explains (in appendix 9 of the report), documented evidence over the last 100 years or so has shown that this is not a sensible approach when teaching students in their pre-university years. Rather, as the report suggests, it is preferable to start with some well-known or 'obvious' facts. These need to be carefully chosen and, in a sense, be explicit. Then, the report goes on, through using deductive reasoning, a collection of related results, of a less obvious nature, can be built up. This is what is sometimes called *local deduction*, where learners can utilise geometrical properties that they already know to deduce or explain other facts or results. The idea is that fluency with local deduction should provide a foundation on which to successfully build knowledge of systematic axiomatisation at a later, and more appropriate, stage of their mathematics education. At such a later stage, facts or theorems taken as 'obvious' at the earlier stage can be revisited with a view to asking how they might be proved, avoiding the problem of infinite regress by invoking a minimal set of initial reasonable assumptions (i.e. axioms). Perks and Prestage (2006), for example, suggest an approach that has isosceles triangles as the basic unit for reasoning. Indeed, the different translators and interpreters of the books of Euclid who have used slightly different sets of axioms are also an example of ways of ordering theorems from what is 'obvious'.

Fluency with local deduction, as Turnau (2002) explains, is best developed through learners' problem solving activity. This includes what Brown *et al.* (2003, p. 6) list as a wide range of activity, including:

Deriving a specific value of a variable (e.g. the size of an angle) using both known theorems and known properties of shapes.

Deducing a specific result in relation to a figure with given properties which does not have the generality or status of a theorem (e.g. proving that two sides of a quadrilateral with a particular set of properties are equal). This type of problem used to be known as a 'rider'.

Considering alternative definitions of geometrical shapes, deciding which of these are necessary, sufficient and minimal, and becoming familiar with the differences between the meanings of these terms.

 www.nuffieldfoundation.
org/geometry-7

 www.nuffieldfoundation.org/
geometry-8

## Using software

Using DGS in teaching provides an environment in which it is possible to create and then manipulate geometric constructions, thereby providing opportunities to interact with geometrical theorems and see the results. A useful comparative review of examples of DGS is provided by Mackrell (2011). Several threads can be identified in research on the use of DGS, including the ways in which students learn to use the capability of the software to 'drag' objects (such as points) when constructing geometrical figures (e.g. Baccaglini-Frank and Mariotti, 2010; Hollebrands, 2007; Hölzl, 1996; Maymon-Erez and Yerushalmy, 2007) and how this relates to the design of teaching activities so that learners come to understand geometrical ideas and concepts (Erfjord, 2011; Laborde, 2001).

As noted above, an important thread in the research concerns the impact of using DGS on the teaching and learning of proof in geometry. Jones (2000b), for example, shows how the mathematical explanations of students using DGS at the early stages of learning to prove can evolve over time to ones that should help provide a foundation on which to build further notions of deductive reasoning. Olivero and Robutti (2007) show how using measuring in DGS can be a powerful tool in supporting the teaching of conjecturing and proving, while at the same time being a complex tool that requires appropriate management and interpretation.

Research on the use with learners of pre-constructed DGS files is showing how such usage can support, or impede, learners' development of geometric conjecturing and reasoning (see, for example, Leung, 2011; Sinclair, 2003; Trgalová *et al.*, 2011). Papers on the impact of the increasing capabilities of such software to import digital pictures, link geometry and algebra, and provide tools for 3D geometry, are emerging (e.g. Accascina and Rogora, 2006; Jackiw and Sinclair, 2009; Pierce and Stacey, 2011); it is likely that research studies will follow in due course.

 www.nuffieldfoundation.org/geometry-9

## Future reform of the geometry curriculum

Using the methods noted above to teach the current geometry curriculum more effectively does not necessarily mean that calls to reform the geometry curriculum and its teaching, as noted in the introduction to this chapter, are likely to subside. A recent heartfelt example is that of Hestenes (2010, p. 38) who argues from the perspective of a practising scientist that not only is the mathematics taught in secondary school 'fragmented, out of date and inefficient' but also that the main problem lies with secondary school geometry. For Hestenes, improving secondary school geometry entails recognising at least the following:

Geometry is the starting place for physical science as well as providing the foundation for mathematical modeling in physics and engineering, including the science of measurement in the real world

Synthetic methods employed in the standard geometry course are centuries out of date; they are computationally and conceptually inferior to modern methods of analytic geometry, so they are only of marginal interest in real-world applications

A reformulation of Euclidean geometry with modern vector methods centered on kinematics of particle and rigid body motions will simplify theorems and proofs, and vastly increase applicability to physics and engineering.

This illustrates how there is every prospect of the debate about geometry and its teaching continuing. Whatever happens in this debate, it is worth concluding this chapter with the following guidelines for successful teaching adapted from Brown *et al.* (2003, p. 8):

- Geometrical situations selected for the classroom should, as far as possible, be chosen to be useful, interesting and/or surprising to students.
- Classroom tasks should expect students to explain, justify, or reason and provide opportunities for them to be critical of their own, and their peers', explanations.
- Tasks should provide opportunities for students to develop problem solving skills and to engage in problem posing.
- The forms of reasoning expected should be examples of *local* deduction, where students can utilise geometrical properties that they know to deduce or explain other facts or results.
- In order to build on learners' prior experience, classroom tasks should involve the properties of 2D and 3D shapes, aspects of position and direction, and the use of transformation-based arguments that are about the geometrical situation being studied (rather than being about the transformations themselves solely as mathematical operations).
- While measures are important in mathematics, and can play a part in the building of conjectures, the generating of data in the form of measurements should not necessarily be an end point to learners' geometrical activity. Indeed, where sensible, and in order to build geometric reasoning and counter possibly deep-seated reliance on empirical verification, it is worth considering classroom tasks where measurements (or other forms of data), or purely perceptual reasoning, are *not* generated.

## Summary

This chapter argues that secondary school geometry is not solely about circle theorems or using Pythagoras' theorem; rather, as Malkevitch (2009, p. 14) illustrates, geometry is more akin to 'the branch of mathematics that studies visual phenomena' in all their glories and richness. This is why geometry is such an important part of the school mathematics curriculum, and why the teaching of geometry across the 9–19 age range needs to ensure a sustained focus on the twinned aspects of geometry across both 2D (plane) and 3D (solid) geometry: the spatial aspects, and the aspects that relate to reasoning with geometrical theory (utilising both transformation and/or congruency arguments). In forming the *yin yang* of geometry education, each gives rise to the other and each only exists in relation to the other.

As Del Grande (1990, p. 19) argued 'geometry has been difficult for students due to an emphasis on the deductive aspects of the subject and a neglect of the underlying spatial abilities acquired by hands-on activities that are necessary prerequisites for understanding and mastery of geometrical concepts'. As this chapter shows, in the curriculum and in its teaching, the twin aspects of geometry, the spatial and the deductive, need to be bound together as inseparable parts of a mutual whole. Only then can learners experience the full power of spatial and geometrical reasoning.

## Where additional evidence is needed

Practitioners are in a position to provide much-needed evidence about the teaching and learning of spatial and geometrical reasoning. Where such additional evidence is needed includes studies that might focus on

- teaching approaches that harness the implicit knowledge of geometry that students bring to lessons in ways that support their learning of spatial and geometrical reasoning;
- ways in which students can utilise geometrical properties that they know to deduce or explain other facts or results that may be less obvious to them as a way of building their capability with local deduction;
- a formulation of the curriculum where interrelationships are more explicit between geometric measurement formulae and how these relate to geometrical theory.

## Key readings

Battista, M. T. (2007). The development of geometric and spatial thinking. In F. Lester (Ed.), *Second handbook of research on mathematics teaching and learning* (pp. 843–908). Reston, VA: NCTM.

This chapter provides an authoritative review of research on spatial and geometrical reasoning.

Bryant, P. (2009). Paper 5: Understanding space and its representation in mathematics. In: *Key understandings in mathematics learning*. London: Nuffield Foundation. http://www.nuffieldfoundation.org/sites/default/files/P5.pdf.

This paper reviews the research on younger children's informal knowledge of spatial relations and their learning of geometrical ideas during primary school. It provides pointers to the spatial and geometrical learning that is possible beyond the age of 9.

Owens, K. and Outhred, L. (2006). The complexity of learning geometry and measurement. In A. Gutiérrez and P. Boero (Eds), *Handbook of research on the psychology of mathematics education: Past, present and future* (pp. 83–115). Rotterdam, Netherlands: Sense Publishers.

This is a useful summary of research into the learning of geometry, including topics in geometrical measurement.

CHAPTER 6

# Reasoning about data

## Introduction

The statistics curriculum faces two fundamental challenges. First, it is argued by some that the natural home for statistics is not in mathematics (Smith, 2004). Certainly, the scope of statistics spreads across the curriculum. Whereas mathematical reasoning tends to be validated by acontextual logic, statistical reasoning draws on context and those contexts relate to most disciplines in the school curriculum. Nevertheless, clear connections with mathematics will emerge from this chapter. For example, variability is a key characteristic of phenomena that students are trying to describe, model, or more broadly comprehend. Variability occurs of course when considering mathematical functions and other entities that might be described algebraically or geometrically. However, variability that might be modelled statistically has added layers of complexity because the variability may (or may not) be attributable to a number of causes. Insofar as statistics is a tool for rooting out possible causes or associations, it is invaluable to the practice of many disciplines, but the fundamental notion of variability and indeed invariance lies also at the heart of mathematics. There is then an imperative for the development of particular ways of reasoning statistically as a variation on mathematical variability as well as a tool for making sense of data in other subject areas or to be used by citizens seeking to be better informed.

Second, recent reform is re-positioning the teaching of statistics as a form of enquiry. Traditionally, it has been typical for statistics to be portrayed in national curricula as a subject where you simply compute numerical and graphical representations of data. This chapter focuses on research related to a proposed shift to an enquiry-based approach. In this respect, this chapter stands in contrast to most other chapters, which focus on how ideas are being taught. In this chapter, we report and critique research studies centred on students' inference making when they work with data arising out of experimentation or observation. First, there is a discussion of the nature of reasoning about data. Since the emphasis on informal inferential reasoning may be unfamiliar to many readers, an illustrative scenario is offered. The next section describes what related knowledge students might bring to their learning around this topic. Then we provide an analysis of students' reasoning about data, followed by a section on some researchers' cognitive models of expert-like statistical reasoning and the curriculum. This analysis makes it possible to discuss possible teaching approaches before presenting a summary and areas in which more evidence is needed.

## The nature of reasoning about data

Variability in data may be explained or modelled as the effect of identifiable factors. Alternatively, variability may be regarded as random noise or error by which it is meant that the variability is not ascribed to any specific factor. How might naïve students make judgements of what should be regarded as variability due to a main effect and what should not? Statisticians have developed methods for making such judgements but research is interested in how, at an informal level, students might make such judgements and how might teachers or resources support more sensitive or sophisticated informal inferential reasoning about underlying trends.

As in earlier chapters, rather than reporting exhaustively on research in this area, the aim is to critique that research in ways that provide salient information to teachers and those working with teachers in initial teacher education or continuing professional development.

## An illustrative scenario

The notion that statistics is primarily an enquiry involving reasoning about data may be unfamiliar to some teachers, as might some of the associated terminology. In this section, we present a fictitious exploration of data in order to introduce the idea. Later, when we discuss the research, we will refer back to this scenario to exemplify some of the research-based ideas.

In fact, this scenario is an imaginative reconstruction of a real episode, briefly described by Shaughnessy (2007, p. 973) when students explored data about the eruption times of the Old Faithful Geyser. The students were asked to make a prediction about how long they would expect to wait for an eruption of Old Faithful. The process described below, though a fantasy in six parts, reflects some of the events described in Shaughnessy's account but is not a faithful (old or not) account of the original episode.

Step 1

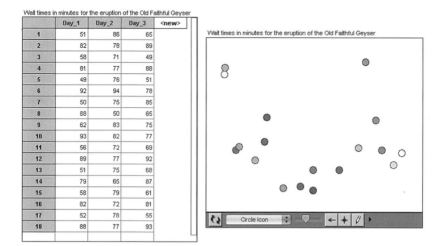

Wait times in minutes for the eruption of the Old Faithful Geyser

| | Day_1 | Day_2 | Day_3 | <new> |
|---|---|---|---|---|
| 1 | 51 | 86 | 65 | |
| 2 | 82 | 78 | 89 | |
| 3 | 58 | 71 | 49 | |
| 4 | 81 | 77 | 88 | |
| 5 | 49 | 76 | 51 | |
| 6 | 92 | 94 | 78 | |
| 7 | 50 | 75 | 85 | |
| 8 | 88 | 50 | 65 | |
| 9 | 62 | 83 | 75 | |
| 10 | 93 | 82 | 77 | |
| 11 | 56 | 72 | 69 | |
| 12 | 89 | 77 | 92 | |
| 13 | 51 | 75 | 68 | |
| 14 | 79 | 65 | 87 | |
| 15 | 58 | 79 | 61 | |
| 16 | 82 | 72 | 81 | |
| 17 | 52 | 78 | 55 | |
| 18 | 88 | 77 | 93 | |

**Figure 6.1** The students have entered the data into TinkerPlots™ (http://www.keycurriculum.com/products/tinkerplots), software that supports the intuitive graphing of data and the easy calculation of statistics. The data shows 18 successive wait times for each of three days. Because the data are not yet structured by any axes, each case appears positioned arbitrarily on the screen in the plot on the right-hand side.

 http://www.keycurriculum.com/products/tinkerplots

## Step 2

Wait times in minutes for the eruption of the Old Faithful Ge...

| | Day_1 | Day_2 | Day_3 | <new> |
|---|---|---|---|---|
| 1 | 51 | 86 | 65 | |
| 2 | 82 | 78 | 89 | |
| 3 | 58 | 71 | 49 | |
| 4 | 81 | 77 | 88 | |
| 5 | 49 | 76 | 51 | |
| 6 | 92 | 94 | 78 | |
| 7 | 50 | 75 | 85 | |
| 8 | 88 | 50 | 65 | |
| 9 | 62 | 83 | 75 | |

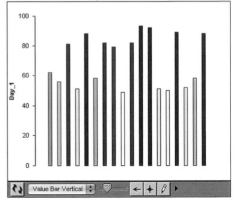

**Figure 6.2** The students drag the Day 1 attribute onto the vertical axis and show each case as a vertical bar. The students notice how the wait times vary. Because the data are not structured on the horizontal axis, it is not possible to draw inferences about any patterns in that variability.

## Step 3

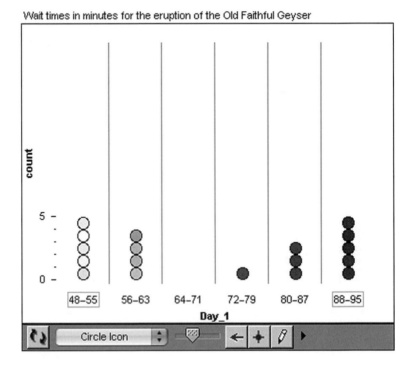

**Figure 6.3** The students randomise the position of the data once more and then drag the Day 1 attribute onto the horizontal axis. The waiting times are grouped into bins of width 8. The students notice how low and high waiting times seem to occur more often than waiting times from the middle of the range.

## Step 4

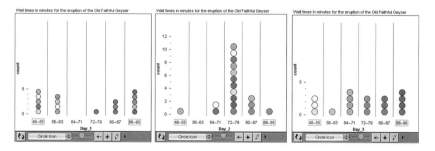

**Figure 6.4** The students repeated this process for Day 2 and Day 3 and found quite different patterns in the data. For example, Day 2 shows data that is much more centralised.

## Step 5

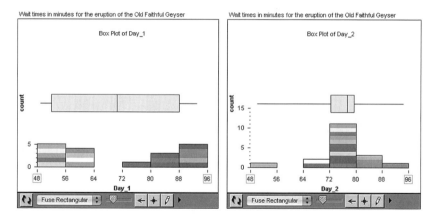

**Figure 6.5** The students created histograms for each of the Days by changing the shape of the icons. They then created a Box Plot, which showed the overall size and spread of the data. The box plots confirmed the data in Day 1 were much more spread than those in Day 2.

Step 6

Wait times in minutes for the eruption of the Old Faithful Geyser

| | Case | Day_1 | Day_2 | Day_3 | <new> |
|---|---|---|---|---|---|
| 1 | 1 | 51 | 86 | 65 | |
| 2 | 2 | 82 | 78 | 89 | |
| 3 | 3 | 58 | 71 | 49 | |
| 4 | 4 | 81 | 77 | 88 | |
| 5 | 5 | 49 | 76 | 51 | |
| 6 | 6 | 92 | 94 | 78 | |
| 7 | 7 | 50 | 75 | 85 | |
| 8 | 8 | 88 | 50 | 65 | |
| 9 | 9 | 62 | 83 | 75 | |
| 10 | 10 | 93 | 82 | 77 | |

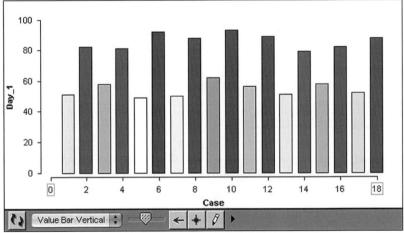

**Figure 6.6** The students have decided to look at how the wait times change over time. They entered a case number into the Case Table so that the first wait time was numbered 1, the second 2 and so on. When they dragged Case onto the horizontal axis and Day 1 onto the vertical axis, they were surprised to see the oscillating nature of the wait times. This overall pattern was apparent also for Day 2 and Day 3. The students wondered why that might be.

## Early understanding

Students may typically have experienced a number of numerical and graphical representations. They may often have practised how to calculate various types of average such as mode, median, and mean. Inevitably, they are likely to confuse the meanings and think of average as the same as mean average. Their experience may have been limited to computing these statistics and so they may have little understanding of when each might be useful and their differing characteristics. They are likely to have experienced various types of tables and graphs are ways of presenting data. Graphical representation will typically include bar charts with frequency on the vertical axis and categorical, individual items, grouped and scaled data on the horizontal axis. Other representations might include pie charts and line graphs, the latter probably with time on the x-axis. Since both these forms demand a range of technical skills, the students' competency may sometimes be fairly low or inconsistent. They may not have met scatter graphs, sometimes referred to as scatter plots, and so will probably not have discussed covariation.

All of these representations are likely to have been taught as techniques and so many students are likely to be unclear about the advantages and disadvantages of the different representations in different situations. These representations are unlikely to have been used for analytical purposes and so may be seen as essentially pictures for communicating data. The students may have had only limited experience of conducting their own experiments and explorations and those they have done may have seemed quite contrived, so appreciation of the full data handling cycle will be limited or non-existent.

## Students' reasoning about data

Over the last century, statisticians have developed a battery of techniques aimed at supporting the sort of judgements mentioned above, namely: (i) identifying possible trends in data to make predictions; (ii) testing whether hypotheses are unlikely to be supported by data; (iii) creating models or descriptions of which factors are associated with variability in data. It is perhaps worth noting that (ii) is contrary to the normal approach to reasoning in which people tend to seek out confirmatory evidence.

Some of these techniques have been taught, especially at higher age levels, though often with limited success, at least if the success criterion is based on deep understanding rather than proficiency in executing the technique. There has been an international shift both in the move to reform statistics education and in the focus of research away from learning about representations and techniques to broad statistical enquiry. This change has been prompted by the ideas of Exploratory Data Analysis (EDA) (Tukey, 1977) and stimulated by the increasing availability of innovative digital tools. The above illustrative scenario offers an example of EDA, marked by the use of different graphical forms in each step and the calculation of statistics, such as those that measure the size and spread of the data. For example, in the box plots in the illustrative scenario (Step 5), the size of the data is represented by the median, the vertical line inside the box itself, and the spread by distance between the lower (25% of the data lie below this statistic) and upper (75% of the data lie below) quartiles. This distance is represented by the length of the box. It is worth noting that this way of looking at data was invented as recently as the 1970s.

 www.nuffieldfoundation.org/data-1

Researchers have become interested in how younger students reason with data and, by implication, whether new pedagogic methods and tools, such as TinkerPlots™, might build on those reasoning processes to facilitate deeper understanding. The notion of informal inferential reasoning (IIR) is now used to capture learning processes whether as a precursor to learning classical inference or as an essential piece of equipment for the modern statistically literate citizen who needs to be able to reason with and about data.

Makar and Rubin (2007) suggest that the conceptual roots of statistical inference lie in 'the process of making probabilistic generalizations from (evidenced with) data that extend beyond the data collected' (p. 2). Building on this broad notion, Zeiffler *et al.* (2008) offer one helpful definition that describes IIR 'as the way in which students use their informal statistical knowledge to make arguments to support inferences about unknown populations based on observed samples' (p. 44). They suggest that IIR includes: (i) making judgements and predictions about populations based on samples; (ii) drawing on prior formal and informal statistical knowledge; (iii) articulating evidence-based justifications. In the above illustrative scenario, one could think about the wait time data as a sample taken across three days of all of the wait times that could be measured over eternity. The challenge to predict an expected wait time is a request to make a judgement about the population based on the sample of data available. Formal knowledge is used throughout the steps in the way that the data is organised and re-organised using tables, stick charts, histograms, box plots, and trend graphs and in the use of summary statistics, such as medians and quartiles. Informal knowledge is used in trying to make sense of that data, such as associating the variability with the observation that geysers are unpredictable in exactly when they will erupt. When the students begin to wonder why the Geyser's wait times oscillate, they are searching for justifications of the apparent pattern observed in Step 6.

There is evidence that students improve with maturation in their understanding of inference. For example, Jane Watson (2001) observed 42 students over three or four years and found that more than half improved in this respect without any specifically designed intervention, though clearly there will have been many influences during those years on those students beyond personal development. Such 'natural' improvement does not, nevertheless, seem to prepare students either for subsequent formal statistical inference or for participation in adult society as might be expected from a statistical literate citizen. Researchers have therefore been exploring students' IIR, focusing on the different elements of the above definition.

 www.nuffieldfoundation.org/data-2

 www.nuffieldfoundation.org/data-4

One emphasis in these definitions is that of relating sample to population. Classroom tasks often involve collecting and presenting data where there is no explicit question driving the collection of that data. Even if there is a purpose to the data collection, a teacher may need to make explicit that the data refer to a sample whereas the question refers to a population. Pratt *et al.* (2008) distinguish between two scenarios, called Game 1 and Game 2. In Game 1, the dataset is all there is; it describes the whole population and there is no reason for inference. In Game 2, a sample has been collected and so there is a demand for inferential reasoning.

Makar and Rubin (2009) quote one of their teachers, Natasha, as arguing that imperatives in task design that make possible generalisation beyond data are: (i) posing a driving question; (ii) including an engaging context; (iii) ensuring sufficient complexity in the data. The same researchers refer to data as evidence being another hallmark of IIR and refer to students who assumed that the handspans in their class would be just like those of a neighbouring class. It was only by collecting other samples that they began to appreciate the variability in the data and therefore the need for tentativeness in making inferences. This tentativeness was expressed as uncertainty and levels of confidence, which Makar and Rubin see as the third key component of IIR. Later, we discuss pedagogic methods of supporting these awarenesses in students.

It cannot be assumed that students will appreciate how larger samples offer less uncertainty. This idea will be explored in more detail in the later section on sampling and also with respect to the Law of Large Numbers in the next chapter but it is worth mentioning one piece of work under this heading of IIR. Pratt *et al.* (2008) investigated 10–11-year-olds' ideas about sample size within an exploration of IIR. They gave the students a simulation of a non-standard die and the students were challenged to infer its unknown configuration from virtual throwing of the die. They could throw the die as many times as they liked. The students however did not always see the value of throwing the die more often as this was sometimes seen as simply complicating the data. The students tended to focus naturally on the variability from throw to throw rather than on the aggregated data. Pratt *et al.* argued that, when attention is locally oriented on the here and now, it is reasonable to think that more data are unhelpful; effective IIR needs to take a global perspective on aggregation in the long-term. The implication here is that unless schools provide the emphasis on aggregation, students will continue to focus only on local variation. More research is needed but it is reasonable to suppose that a tendency not to think about aggregates in the long term might be related to a tendency not to think strategically. A preference instead to focus on immediate outcomes has been reported elsewhere (see Konold's *outcome approach* in the next chapter) and seems to be an ongoing challenge for older students, including adults.

The whole of this section on students' reasoning assumes a focus on IIR. However, there are a number of sub-strands in that research that can be identified and each of these will become a focus for a subsection.

## Reasoning about variability in data

In response to calls (Shaughnessy, 1997, amongst others), research on variability has blossomed in the last decade. Shaughnessy (2007) returned to this theme by alerting educators to the dangers of focusing too quickly on measures of position (average) without due consideration of variability. His students analysed data for three consecutive waiting times in minutes for the next eruption of the Old Faithful Geyser (see the illustrative scenario). The rush to calculate the mean average disguised the cyclic nature of the waiting times.

www.nuffieldfoundation.org/data-1

Pfannkuch (2005), in discussing a range of papers on students' reasoning about variability, which appeared in a special issue of *Statistics Education Research Journal,* identified a contrast between student activity that is focused on pre-classical inferential competences (testing hypotheses) and more general modelling activity as might be conducted by a statistician who is aiming to model variation (seeking trends). Perhaps Shaughnessy's Old Faithful example illustrates the need not only to test conjectures but also to look to model variability. Moving quickly to calculation of average assumes that the variability can be discounted, whereas an exploratory data analysis approach might highlight the significance of the variability as eventually was apparent in Step 7 in the illustrative scenario.

Ben-Zvi (2004) studied 13-year-old students in an experimental school as they sought to make comparisons between the lengths of English-American and Hebrew names. By studying two students in detail, Ben-Zvi identified a number of developmental steps in how the students harnessed variability. These moved through stages of:

- where to focus attention;
- describing variability;
- conjecturing possible explanations for that variability;
- measuring variability;
- modelling variability (in particular managing outliers);
- finally noticing variability within and between distributions.

Although these steps are described as developmental, they are closely related to the teaching sequence and so it may be safer to regard them as distinctive aspects of the pedagogic process, as is acknowledged by the author. Indeed, Ben-Zvi recognised the significance of decisions that the students should be encouraged to experiment repeatedly with tools, draw on previous experience, integrate contextual and statistical knowledge, and benefit from the power of technological tools and interactions with the teacher.

www.nuffieldfoundation.org/data-3

## Reasoning about variability in samples

There are various types of relationship in Game 2 (Pratt *et al.*, 2008, as mentioned above) between the sample and the population. Compare for example a sample used to draw inferences in market research about a population's needs or desires, where the population is finite but large, with a sample of coloured sweets taken from a bag where the sample is not only finite but the population itself is easily counted. The purpose of asking a sample of people their opinions in market research is self-evidently sensible as it would be impossible or very costly to ask everyone in the population. The market research example is clearly Game 2 whereas the sweets in a bag example perhaps should be Game 1 (after all, it would be fairly simple to remove the all of the sweets and count them!) and is artificially made to look like Game 2 for pedagogic reasons. A third distinctive case of the relationship between sample and population is the commonly used sample of totals of two dice. Here, the population might be envisaged as a theoretical probability distribution (each possible total with an associated probability) or as an infinite set of totals that might be generated in the future. Either way, the population does not possess the material substance of people or sweets. Nevertheless, in all three examples, the challenge is to draw a conclusion about the population, or some aspect of the population based on the evidence in the sample.

In the illustrative scenario, the students might have been challenged to find the mean average wait time from the data available. This would then be a task simply to describe the data (Game 1). Since they were asked to find how long they might expect to wait (in the future), the students needed to consider not just this data but all possible wait times (Game 2). Indeed, just as the total of two dice can be thought of as being a random variable with a probability distribution, statisticians might think of the population of wait times as a probability distribution and the values that happen to be observed to have been generated randomly from that distribution. Initially, the statistician might, for example, think that a Poisson distribution would offer a good model of the population of wait times. Poisson distributions tend to be relevant when the number of occurrences of a randomly occurring phenomenon in a finite time is being counted and when the mean average rate of occurrence is constant over the whole duration. However, if they went through a similar EDA process as the imagined students, they would have discovered that the average wait time for the eruptions did not seem to be invariant as evidenced by the differences between the three days and the oscillating pattern in Step 6. As a result, the statistician may reject the Poisson model to consider other options. So, a population might consist of material cases, like the people whose opinions might have been sought in a piece of research, or mathematical/statistical functions, like the total of two dice or the distribution of wait times. The sample might be partial out of necessity as in market research, out of pedagogic design, as in the sweets in a jar, or part of an infinite population, as in the case of theoretical distributions.

Variability in samples, the title of this subsection, has intentional ambiguity. The interpretation might be variability *between* samples; much of the research on reasoning about sampling distributions has related to university-level education, where the concept becomes a major focus of instruction. At school level, much of the research has focused on variability *within* a sample. Such variability is tightly connected to distributional thinking, which is a focus of the next chapter.

 www.nuffieldfoundation.org/data-3

Limited research has so far been conducted in how school students conceive of a sample. Watson and Kelly (2005) conducted large-scale research across 639 students at a school. The students were pre-tested, post-tested, and delay post-tested, in some cases after specific instruction. There was a chronological improvement in understanding of sampling but there was no significant difference between those who received the specific instruction and those who did not. In another study, Watson and Moritz (2000) researched ideas about sampling amongst 62 students aged between 8 and 15 years. They adopted Biggs and Collis's (1982) SOLO (Structure of Observed Learning Outcomes) model to identify distinct types of responses to their questionnaire.[1] At the simplest level (referred to in

---

[1] More detail about the SOLO approach is given in the next chapter.

SOLO as *unistructural* because the student describes only one aspect of the concept), students described a sample as 'a little bit' or 'something to test'. At a more complex level (called *multistructural* because the student refers to two or more aspects of the concept), students referred to the population as a reference point for the sample. At a yet more advanced level (*relational*), students referred to not only the piece (the sample) and the whole (the population) but also the testing aspect (whether the sample was large enough to make inferences).

Echoes of these responses can be found in one of two types of thinking about sampling in the research by Saldanha and Thompson (2002) on 16/17-year-old students. Their approach was to use news reports that mentioned data about sampled populations and to interrogate the students about the likely frequency of such results if the experiment were repeated. Large numbers of students conceived of the sample as a simple subset of the population, a construction referred to as *additive*, which appears to be closely related to the responses of Watson and Moritz's students, perhaps at all levels. Some students were able to construct an image of sampling as *multiplicative*, in the sense that the resemblance between sample and population was blurred by ideas of variability in distribution of a collection of sample proportions. These students perhaps benefitted from the re-sampling methods that exploited modern software and were not available to the students in the Watson and Moritz study.

www.nuffieldfoundation.org/data-5

Re-sampling is a process made accessible by digital technology and it potentially helps to direct attention to how, in Game 2, when a sample is taken, it is a random selection of cases such that if the population were re-sampled the new sample would not be identical to the previous one. Hence there comes a focus on variability *between* samples. Furthermore, if a statistic such as the sample mean were calculated for the second sample, it would be likely to differ from that in the first sample. If a large number of samples were taken and the sample mean (or any other statistic) were calculated each time, it becomes possible to look at a sample of sample means, taken from an imagined population of sample means.

In re-sampling, there is a need for a large number of samples to produce a reasonably clear image of the sampling distribution; it is not really feasible to execute re-sampling by hand. Re-sampling is illustrated using version 2 of TinkerPlots™ in Figure 6.7. In the top left hand corner, two spinners have been set up to simulate standard dice. The table to its right shows a sample of ten throws generated by this simulation. The total of the two dice scores has been calculated. To the right of the table is a plot of those ten throws, showing also the mean of that sample as a small triangle on the horizontal axis. Each time the simulation is run, the values in the table vary as do the plot and the sample mean. In the bottom left-hand corner, a history is kept of sample means. In all, a sample of 100 samples means is collected by running the simulation 100 times. The histogram in the bottom right-hand corner displays those sample means and also indicates the mean of the sample means as a small triangle on the horizontal axis. The histogram gives some indication of what the population of sample means might look like and the mean of the sample means gives some indication of what the mean of the population is.

Saldanha and Thompson were surprised at the frequency with which the simpler additive construction of sample was articulated despite focused teaching using re-sampling techniques. They suspected that the difficulty for these students lies in the complexity of re-sampling, where it was common to confuse the number of samples with the number of cases in the original data. For example, there might be confusion between the cases in the sample of throws in the top row of the figure and the cases in the sample of sample means in the bottom row.

Despite this result and the apparent ineffectiveness of the teaching in the study by Watson and Kelly, Wild *et al.* (2011) remain optimistic that design solutions are possible that will resolve the confusions in re-sampling, pinning their hopes on the use of visual comparisons to enable inferential reasoning to remain focused on the relevant graphs of data. It remains to be seen whether such novel approaches will resolve the sort of confusions apparent in the reasoning of Saldanha and Thompson's students. We say more about this in the pedagogy section later.

www.nuffieldfoundation.org/data-9

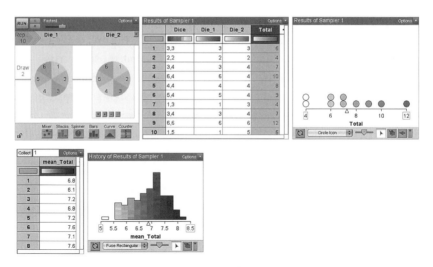

**Figure 6.7** Re-sampling in TinkerPlots™ from the total of two dice.

Pfannkuch (2005) emphasised the crucial nature of the full enquiry cycle, pointing out that often research (and indeed teaching) focuses on specific aspects of that cycle. For example, if data are provided to the student, it needs to be recognised that there may be a shortage of information about how the data were collected and about the context. She is not arguing that it is therefore essential that students always collect their own data but rather that, without such information, reasoning about variability (whether in data or in samples) is likely to be limited. The validity of this observation is perhaps emphasised by the research on the role of context.

## The role of context in informal inferential reasoning

More recently, researchers have been examining the role of context in IIR. In trying to understand the role of context in statistical enquiry, let us first examine the corresponding research in mathematics education. Research in mathematics education has reported how the context in word problems can act as an obstacle obscuring access to the mathematical ideas embedded in the problem (Cooper and Dunne, 2000). Paradoxically, other researchers have also reported how the setting provides meaningfulness that affords a facility and flexibility not apparent when the mathematics is learned out of authentic contexts in the classroom (Nunes *et al.*, 1993). One might characterise this tension as the 'context dilemma': on the one hand, teachers may want to present the mathematics in context in order to engender meaningfulness, but on the other hand they may not want the context to distract from the mathematics.

It could be argued that the role of context in statistical enquiry is fundamentally different from that in mathematics education. Whereas mathematics in its purest sense derives its rigour from the denial of situated factors, statistics as an applied subject needs to draw conclusions about hypotheses set in pragmatic contexts. So, one might expect the context dilemma to be less apparent in IIR. Nevertheless, Dierdorp *et al.* (2011) pointed to the importance of authentic practices in supporting students IIR about correlated data. In particular, they attached importance to the students collecting their own data to promote familiarity with the context. Langrall *et al.* (2011) have also reported on the role of context familiarity when students compare data. Eighteen middle school students analysed authentic data relating to their personal areas of interest. They found that these students used context knowledge to inform the enquiry and to explain, justify, and qualify their conclusions, thus recognising the importance of context in supporting meaning making. Gill and Ben-Zvi (2011) discussed how the explanations of their 12-year-old students served a variety of purposes, but always seemed to involve the intertwining of contextual and statistical considerations. In contrast, Pfannkuch (2011) offered a complex picture from her research with ten 14-year-olds. She found that contextual knowledge did assist students in finding meaning from observed patterns but could also divert their attention during the construction of concepts, a re-statement in effect of the context dilemma above. So it seems that context is seen as both positive and negative in statistics education as well as in mathematics education more generally.

Perhaps the resolution of the context dilemma may be forthcoming if close attention is paid to the agenda apparent in the task design in these different situations. In mathematics education research, a focus on the mathematical representations or operations seemed to lead to a perspective in which context detracts from the construction of meaning about that piece of mathematics. Similarly, the research in statistics education can more easily recognise the benefits of context when the focus is on statistical enquiry than when it is on teaching a particular concept. More insight into this difference can be gained by reference to an aspect of the philosophical literature which focuses on the nature of knowledge.

Some researchers, such as Bakker and Derry (2011), argue that statistical knowledge is set in the contextual domain, justifying this position through the contemporary philosophical stance called *inferentialism* (Brandom, 2002) in which reasons for knowing are privileged over representations of knowledge. That is to say that people invent representations because they are useful and that knowledge is fundamentally laid out not in terms of representations but in relation to the contextual connections that drive the need to know. From this position, curricula and syllabi that list statistical knowledge in terms of representation such as types of graphs (for example, bar charts, pie charts, histograms) and measures (of position or dispersion, for example) misunderstand the nature of that knowledge and leave teachers the difficult challenge of re-formulating the syllabus so that those representations are encountered during meaningful tasks and project work. Ainley *et al.* (2006) have described that challenge in terms of connecting *purpose* to *utility* in task design. The inferentialist position has implications for pedagogic approach, which will be discussed later. Contrast, for example, the aim of teaching that focuses on the calculation of different representations of average with the purpose of enquiry exemplified in the illustrative scenario above, which puts to the foreground reasons for engagement with whatever representations might be needed to further the enquiry.

www.nuffieldfoundation.org/data-7

IIR is taken to operate as a broad perspective on how reasoning about data takes place. Makar *et al.* (2011), in discussing the role of context in IIR, emphasised the need for an enquiry-based environment in which conflicts between what students believe and what the data suggests drive the search for explanations. For example, two boys in their research focused on data about how far students in Grade 6 and 7 were able to jump. They expected the older children to jump further but the data showed that the Grade 6 children tended to jump further. The surprise in this conflict led them to conjecture that there might be a gender-based explanation as they expected boys to jump further than girls. In fact, they found only one girl in their Grade 6 sample.

In addition though, they reported how their 12-year-old students were also dependent on specific statistical knowledge of concepts and tools to deepen their emerging understandings. Their use of graphs and averages was essential in confirming or denying their conjectures throughout the exploration. Even an inferentialist stance on IIR acknowledges the special role played by specific knowledge about particular representations of statistical concepts or families of concepts that will become relevant in the search for meaning. In the next two subsections, we examine research first on representations of signal and noise and second on graphical representations.

## Reasoning about representations of signal and noise

Konold and Pollatsek (2002) have proposed that the key organising idea in the study of statistics as an exercise in Exploratory Data Analysis should be the study of noisy processes within which can be detected a signal given sufficient output. These authors criticise curriculum materials as portraying averages as mere summaries of groups of values, in contrast to the statistician's perspective on averages as central tendencies or data as 'a combination of signal and noise' (p. 261). Other meanings for average are, in their analysis: typical value, fair share, and data reducer. A focus, however, on signal and noise commits one to think about the distribution of data around a signal, so that signal plays a key role, in contrast say to a view in which the mean (or other measures of average such as mode and median) is simply one of many parameters of the distribution. All of these ideas about average are of course true but offer different perspectives. So, in simple situations, it is possible to consider the signal as the average and the noise as the variability around that average but in more complicated situations the signal might not be about central tendency.

In the illustrative scenario, the students are searching for a signal that will enable them to predict the next wait time. No such signal can easily be identified, as there appears to be a great deal of variability in both the average and spread in the early steps. When the data were organised chronologically in Step 6, it becomes apparent that there is a signal of cyclic oscillation. The fact that the data do not conform exactly to any such signal is shown by the variability from oscillation to oscillation, which might therefore be regarded as error or noise. In this sense, a perspective on signal and noise is more general than say a focus on average (also known as central tendency and position) and spread (dispersion).

At the same time, there is a commitment to an inferentialist exploration of data in the pursuit of seeking out the signal and to key concepts rather than

www.nuffieldfoundation.org/data-1

on specific measures of those concepts. For example, conventional approaches often focus on calculations: mean, mode, median, geometric mean, and so on in the case of average; range, inter-quartile range, standard deviation, and so on in the case of spread.

Although there is little evidence yet about whether a focus on signal and noise would be effective, there is research that shows the ineffectiveness of conventional methods. There is, for example, a tendency to always refer to the mean when trying to summarise data. For example, Pollatsek *et al.* (1981) demonstrated how college students would use the mean to represent a set of data in which there was an obvious outlier, which was ignored. Such views probably emerge from a focus on calculation of average and it is unclear whether the essential change in approach should be towards signal and noise or conceptual aspects of average. Shaughnessy (2007) in his review argued that a signal and noise perspective might be too difficult for young students especially since it requires recognition that there is such a thing as noise in data and there are many causes of noise; instead he proposes that a focus on average as fair share or typical value relates more closely to naïve intuitions. However, remember that early research on average mentioned above was at a time when students did not have access to technological support that allows data to be manipulated and represented such as in the illustrative scenario. Such tools might make signal and noise accessible even to young students and certainly it is easier to embed such an approach in an enquiry-based exploration of data, where the aim is to seek out the signal amid the noise, rather like a data detective.

www.nuffieldfoundation.org/data-8

# Reasoning about graphical representations

One of the key affordances of such technological support is to display data quickly and easily in graphical form. Although much research has been done on the difficulties in constructing graphs with paper and pencil (see Chapter 8), by reducing the skill threshold in creating axes and plotting points, technology provides the opportunity to focus on interpretation of graphs, which arguably is a more important skill, at least in terms of statistical literacy.

Graphical interpretation involves understanding graphs created by others, such as reading graphs in the media, and those created by the interpreter through button clicks and menu choices in the software. Both aspects of interpretation seem to involve at least three components (Curcio, 1989): reading the graph; reading within the graph; reading beyond the graph. Reading the graph, such as identifying or plotting particular points against the scales of the axes, is not trivial but is perhaps the most elementary component. Reading within the graph involves understanding the general trend and interpreting particular features such as peaks or troughs within the data. Typically, reading within the graph will be articulated in terms of the graph itself and extend to its wider meaning. Reading beyond the graph involves relating the features of the graph to its wider context, which might involve some projection. A fourth component, reading behind the data, has been proposed (Shaughnessy *et al.*, 1996) to include those aspects of statistical literacy that raise questions about the sources of data, the sampling methods or the population from which the data were drawn. These components of graphical interpretation can be associated with the three tiers of statistical literacy apparent when making sense of media graphs (Watson and Moritz, 1997): knowledge of basic statistical terminology; applying statistics in context; and challenging statistical claims.

The idea that a focus on particular data points is where students start has been supported by other research (Chick and Watson, 2001; Cobb, 1999; Konold and Higgins, 2003) and has led to a broad recognition that methods of displaying data need at least in the early stages to maintain the cases in the display. Note how, in the illustrative scenario, the data is initially displayed as a set of cases in the table (Step 1). Transitions to graphical forms that hide individual cases such as bar charts and histogram (as in Steps 4 and 5 in the illustrative scenario) in order to focus on aggregated data need to be planned carefully. Konold and Higgins (2003), for example, propose the use of a value bar (Step 2 in the illustrative scenario), before adopting the end point of the value bar and stacked dot plots (Step 3).

www.nuffieldfoundation.
org/data-2

 www.nuffieldfoundation.org/
data-6

The focus in this subsection has been on graphs of data collected through survey, observation, or experiment. In contrast to graphs of mathematical functions, graphs of data reflect the messiness of that data in the sense that patterns and trends in the data are implicit and need to be inferred, whereas graphs of functions are rather more explicit and directly accessible. Nevertheless, there is a strong connection between graphing functions and graphing data (which incidentally means that much of the very extensive research in mathematics education on graphing of functions is relevant here but is beyond the scope of this chapter). Consider, for example, tasks involving measurement, such as when teachers might ask students to 'discover' π by measuring the diameters and circumferences of circles. Such experiments inevitably involve inaccuracies so that data with noise, as in the measurements, have embedded within them a signal, in this case the ratio between the circumference and the diameter.

The connection between graphing functions and data has been exploited by some researchers. Ainley *et al.* (2000) have shown how students can gain a sense of the analytical utility of a scattergraph by deploying a particular pedagogic approach called *active graphing*. Students conduct experiments involving bivariate data. Two examples are designing spinners to find the wing length that gives the longest time of flight and rolling cars down a slope to explore the relationship between weight and distance. The students generate scattergraphs from spreadsheets even as they are collecting the data. The scattergraph becomes a tool for deciding what to do next in the experiment. Interestingly, the technique can work even when a mathematical function might describe the relationship.

Thus, the well-known sheep pen problem might be transformed. In this task, the students have to maximise the area of a rectangle where the fourth side of the rectangle is a long wall and a fixed amount of fencing is available to complete the three other sides. The experiment can be conducted practically generating messy data of a statistical kind. When the students recognise that there is an algebraic function that could be used to determine the length of the pen given any particular width, this can be taught to the spreadsheet. At this point, the scattergraph takes on a new appearance in which the data now follow an ideal parabola and the original messy data fits around that curve. The

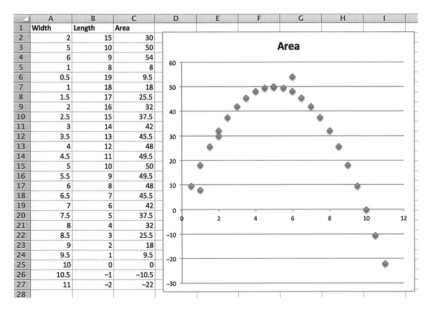

| | A | B | C |
|---|---|---|---|
| 1 | Width | Length | Area |
| 2 | 2 | 15 | 30 |
| 3 | 5 | 10 | 50 |
| 4 | 6 | 9 | 54 |
| 5 | 1 | 8 | 8 |
| 6 | 0.5 | 19 | 9.5 |
| 7 | 1 | 18 | 18 |
| 8 | 1.5 | 17 | 25.5 |
| 9 | 2 | 16 | 32 |
| 10 | 2.5 | 15 | 37.5 |
| 11 | 3 | 14 | 42 |
| 12 | 3.5 | 13 | 45.5 |
| 13 | 4 | 12 | 48 |
| 14 | 4.5 | 11 | 49.5 |
| 15 | 5 | 10 | 50 |
| 16 | 5.5 | 9 | 49.5 |
| 17 | 6 | 8 | 48 |
| 18 | 6.5 | 7 | 45.5 |
| 19 | 7 | 6 | 42 |
| 20 | 7.5 | 5 | 37.5 |
| 21 | 8 | 4 | 32 |
| 22 | 8.5 | 3 | 25.5 |
| 23 | 9 | 2 | 18 |
| 24 | 9.5 | 1 | 9.5 |
| 25 | 10 | 0 | 0 |
| 26 | 10.5 | −1 | −10.5 |
| 27 | 11 | −2 | −22 |
| 28 | | | |

**Figure 6.8** Some of the data in the graph are plotted after manual measurement of the lengths. Most of the data are plotted using a formula for length in terms of width.

connection between the graphing of mathematical functions and experimental data is self-evident as in Figure 6.8.

# Progression in reasoning about data

In this section, we discuss two types of research, the development of normative models of statistical reasoning and arguments about the curriculum.

## Normative models of statistical thinking, reasoning, and literacy

Models that describe students' reasoning have been mentioned above. For example, the types of graphical interpretation: reading the graph and reading within, beyond, or behind the graph. Another example is the use of the SOLO taxonomy to describe how students make sense of sampling.

In this subsection, we focus on normative models, by which we mean descriptions of what is regarded as 'correct' statistical reasoning. Perhaps one of the most helpful normative models has been that by Wild and Pfannkuch

(1999), based primarily on how statisticians describe their work but also correlated with students' work. They proposed four dimensions: the investigative cycle; types of thinking; the interrogative cycle; dispositions. We will give a flavour of each.

The *investigative cycle* moves through identifying the system dynamics of the problem, designing how to measure the identified variables, collecting the data, analysing the data including generating hypotheses, and finally interpreting in order to communicate conclusions.

*Types of thinking* include both the general and the fundamental. General thinking ranges across strategic planning of the problem, seeking explanations, constructing and using models, and applying techniques including problem-solving tools. Fundamental thinking includes recognising that the problem could be informed by data collection, consideration of variability, aggregate-based reasoning with statistical models, drawing appropriately on contextual information, and so-called *transnumeration*. The latter needs some further explanation. Exploratory data analysis involves much data manipulation, trying different graphical representations and numerical measures in search of insights into the stories locked in the data. This process of transforming how the data is presented in an attempt to construct meaning was named by Wild and Pfannkuch as 'transnumeration', their own invented term (p. 227). In the illustrative scenario, the insight gained by creating the trend graph in Step 6 is a good example of transnumeration.

The *interrogative cycle* moves through the generation of plans on how to approach the problem, seeking information and ideas, interpreting in particular by comparing and connecting representations, criticising those interpretations by reference to other sources, and judging what to believe and what to discard.

Finally, *dispositions* range across scepticism, imagination, curiosity, openness, a propensity to seek deeper meanings, being logical, being engaged, and perseverance.

Clearly these dispositions apply equally well to other modes of reasoning than the statistical, as do the general types of thinking. Similarly the investigative and interrogative cycles match other types of modelling and problem solving. In the fundamental types of thinking lies what is distinctively statistical but to focus only on this to the exclusion of the other dimensions would fail to capture the holistic nature of normative statistical reasoning.

# Curriculum content

In some countries, it is becoming established that statistics needs to be taught with a focus on statistical enquiry as reflected in Wild and Pfannkuch's (1999) normative model of statistical reasoning and in the illustrative scenario. Of course, at times there will need to be special attention on particular measures and representations but in many countries, and certainly in the UK, the focus in curricula on representations continues to be prevalent, thus failing to capture much of what it is that statisticians do, and arguably therefore failing to communicate the nature of statistical enquiry.

Pfannkuch (2005) argued that the focus on conducting statistical tests to determine information about the population from the sample data in current curricula discouraged seeking explanations and making predictions. Informal inference might be seen as either a pre-cursor to formal inference or as enculturating young people into a statistical way of thinking. One might have thought that EDA approaches, by encouraging informal inferential reasoning, might move the curriculum away from a tight focus on conducting statistical tests. However, Pfannkuch articulated a concern that insufficient attention was being paid to seeking causal explanations with the result that the exploratory approach seemed more like a pre-cursor to formal statistical inference. If she is right then there is reason to be worried; whereas all students will subsequently need the broader skills of statistical literacy inherent in a statistical way of thinking, only a minority will require an understanding of formal statistical inference.

What does all this mean? In short, a radical shift is proposed in which the curriculum is re-designed to move away from representations and measures and instead embraces statistical enquiry in its fullest sense. This is consistent with the inferentialist philosophy discussed earlier in which representations are seen as important but subsidiary to the reasons for human activity. Even so, according to Pfannkuch, teachers need to be aware of the needs of their students and to place to the fore of statistical enquiry the search for causal explanations, while accepting that not all variation is explainable. In the end, what is being promoted is a modelling approach in which the most prominent factors that cause variation are identified and the rest of the variation is regarded as random. Pfannkuch recognises the need in statistics classrooms for both aspects of informal approaches: those that lead to formal statistical inference and those that focus on the search for causal explanations.

# Teaching approaches to reasoning about data

The previous section has focused on the curriculum in relation to students as learners. In this section, we extract some of the key lessons about teaching.

## Focusing on variability and inference

We described the importance of distinguishing between Game 1 and Game 2 (Pratt *et al.*, 2008) so that the teacher and the students are aware that the aim in Game 2 is to make inferences about the population from the sample of data available. In fact, Wild *et al.* (2011) have adapted this terminology to refer to the *description game* and the *inference game*. Game 2 is often conducted as Exploratory Data Analysis (EDA) (Tukey, 1997). EDA will not happen in classrooms unless curricula encourage it and statistics teachers are prepared to adopt a statistical enquiry approach. Students need to be provided with tools such as TinkerPlots™ that make the approach feasible without an impossible threshold of calculation and graphing to overcome. The investigation involved needs to range across the full enquiry cycle including providing or eliciting information about the sampling process and the population to which the sample refers.

## Focusing on sampling

Two main teaching approaches emerge from the literature. The first has been mentioned above. The notion of variability can be reinforced through the use of software that facilitates re-sampling so that, for example, it becomes clear that the sample mean changes its value in each sample. Re-sampling techniques might enable students to appreciate the extent of the variability in the sample statistic (such as the sample mean) compared to that in the population (for example, the mean).

Wild *et al.* (2011) have argued that the complexities of re-sampling, such as the need to have simultaneously in mind the population, the current sample, and the collection of sampling statistics (see the multiplicity of samples and graphs in Figure 6.7), can be overcome through the use of visual comparisons that enable inferential reasoning without taking the eyes off relevant graphs of data. They aim to place emphasis on the extent to which the sample misrepresents the population rather than just on the variation from sample to sample. To this end, they propose a metaphor that looking at the world by using data is like looking through a window with ripples in the glass. The distorting effect is less when the samples are large. To avoid taking the eyes off the graph, Wild *et al.* are developing visual representations in which box plots of the sample are superimposed at each iteration of re-sampling with the effect that the box plot appears to vibrate or shimmer. The vibration in effect animates the uncertainty in using the sample to tell a story about the population. A single sample is equivalent to a single frame from the vibrating movie.

Although this work is cutting edge at the time of writing this chapter, we believe the ideas have some potential and look forward to seeing how well in practice it supports notions of sampling and sampling distribution.

The second teaching approach that places emphasis on sampling is that instigated by Bakker and described in Bakker and Gravemeijer (2004), referred to as *growing samples*. The approach is to start by asking each student to collect data about themselves and one or two friends. The challenge is whether these data give them any information about the class. When the class data are collected, the challenge becomes whether these data have any value in drawing conclusions about a larger set such as the collection of all students in the year. This process can be continued to the school and beyond. Throughout, the enquiry is focused on to what extent the sample offers information about the population, though the sample grows and so in a sense does the population. Bakker and Gravemeijer report increasing insights about distribution alongside promoting understanding about the uncertainties in sampling. Similar approaches have since been tried with supportive results (Ben-Zvi, 2006).

## Designing purposeful tasks

The research reported in the previous section emphasised the importance of designing purposeful tasks. Makar and Rubin (2009) quoted one of their teachers who emphasised the need for a driving question and complex data. Ainley *et al.* (2000) have described how an active graphing approach supported young students (ages 8 to 12 years) in gaining a sense of the purpose for scattergraphs as having utility in the analysis of data. More generally the focus on enquiry in EDA can only be maintained when the students are pursuing meaningful tasks that cause them to engage with the whole exploratory cycle. Of course, at times teachers need to help students to develop skills and knowledge about specific representations, but the balance whereby at the moment this is the major effort in classrooms needs to change so that knowledge about statistical representations is situated in the context of statistical investigation.

# Summary

Normative statistical reasoning includes developing a mindset in which it is possible to recognise when a problem is amenable to solutions that emerge out of data, in contrast, for example, to problems that are amenable to a numerical or algebraic treatment. That data will inevitably involve variability, not just the sort of variation seen in mathematical functions, but variability which can be partially accounted for by causal factors (or signals) and some which cannot and might be described as random (or noise). The distinguishing of signals from noise might arise out of formal statistical inferential techniques as might be studied at A Level and beyond, but informal progress can be made through EDA, using modern software tools.

Such approaches require informal inferential reasoning, which is not only valuable as a pre-cursor to formal inference but also as a statistical literacy vital for citizens of the future. IIR requires aggregate-based reasoning so that it is possible not only to read within the data but also through and beyond it. Attention must be paid to the full enquiry cycle so that questions about the population and the sample are raised, aspects that might be referred to as reading behind the data. Insights are likely to be gained through transnumeration, the representing of data in a variety of ways until a story within the data is unlocked.

Sampling is a key idea in statistical reasoning and is far from straightforward for students to comprehend. Promise is shown in recent developments in 'growing samples' and techniques geared towards reducing the complexity inherent in re-sampling. Distribution is a related key idea but, since this and sampling are strongly connected to probability, both these ideas are further developed in the next chapter.

# Where additional evidence is needed

Many aspects of reasoning about data have not been thoroughly researched, largely because statistics is a relatively new science and modern digital tools that are impacting fundamentally on the teaching and learning of statistics are relatively novel. Evidence such as that which can be provided by practitioners is much needed.

- The distinction between Game 1 and Game 2 seems fundamentally important. Perhaps explicitly distinguishing between Game 1 and Game 2 with students would be effective strategy. For example, students might first be introduced to the data in a whole population in response to a question such as 'Are boys in this class taller than girls'? This question can be answered definitively, assuming an agreed measure such as the mean for comparing the groups. Then, that situation could be compared to one arising from the question 'Are 11 year old boys taller than 11 year old girls'? This Game 2 question demands consideration of the whole enquiry cycle and provokes questions about whether the data in the sample provides a clear indication of the population. A study, which tracked students' understanding in the Game 2 situation when they have either already explored Game 1 or not, could yield interesting results. The suggestion is reminiscent of the growing samples idea. However, much of that research has been conducted with younger pupils so it would be interesting to find out how well it works with older students.

- Given the debate about the position in the curriculum of statistics, a study might map and compare how statistics is used in different subjects, such as geography, science and PE, where data are used and could be exploited as sources of statistical enquiry, either in the teaching of those subjects or in mathematics classrooms.

- At the moment, re-sampling is perhaps best explored through Fathom™ (http://www.chartwellyorke.com/fathom.html). This is more powerful, more complex software than TinkerPlots™, though it is worth keep a watchful eye on developments by Wild (see http://www.stat.auckland.ac.nz/~wild/VIT/). However, many of the claims for re-sampling are aspirational and not yet rigorously tested. Evidence from projects that explore how students understand re-sampling and whether it can form the basis of an effective teaching strategy is needed.

http://www.chartwellyorke.com/fathom.html

http://www.stat.auckland.ac.nz/~wild/VIT/

## Key readings

Garfield, J. and Ben-Zvi, D. (Eds) (2005). *The challenge of developing statistical literacy, reasoning and thinking*. Dordrecht, The Netherlands: Kluwer.

With chapters from some of the most deep thinking international statistics educators, this edited book provides an excellent basis for the reader who wishes to understand the direction of travel of research in statistics education.

Konold, C. and Higgins, T. L. (2003). Reasoning about data. In J. Kilpatrick, W. G. Martin, and D. Schifter (Eds), *A research companion to principles and standards for school mathematics* (pp. 193–215). Reston, VA: NCTM.

This chapter provides a comprehensive review of research on data analysis at a time when the interest in informal inferential reasoning was beginning to take off. The chapter therefore emphasises data analysis, as opposed to formal inference, but the treatment of the subject is consistent with the approach to IIR taken in Chapter 6.

Shaughnessy, J. M. (2007). Research on statistics learning and reasoning. In F. K. Lester (Ed.), *Second handbook of research on mathematics teaching and learning* (pp. 957–1009). Charlotte, NC: Information Age Publishing.

This is another chapter in a review that embraces all mathematics education. Four years later than that by Konold and Higgins, it provides a very complete summary. In particular it may interest the reader that it includes the research on students' exploration of the eruption times of Old Faithful used in the illustrative scenario. This review argues with the focus on signal and noise proposed by Konold.

CHAPTER 7

# Reasoning about uncertainty

## Introduction

In Chapter 6, we reported and critiqued research studies centred on students' inference-making from data arising from experimental and observational studies. In these situations, variation becomes self-evident in the data and the challenge for the student is to identify underlying trends *despite* that variation. In this sense, the variation in the data is an irritant that would be avoided if only it were possible to eliminate measurement error or have complete data about the population.

In contrast, this chapter examines research studies where students are encouraged to embrace randomness and use it as a means of modelling the uncertainty itself. As in earlier chapters, rather than reporting exhaustively on research in this area, the aim is to critique that research in ways that provide salient information to teachers and those working with teachers in initial teacher education or continuing professional development.

The chapter begins by considering the nature of reasoning about uncertainty by contrasting it with deterministic thinking. In most chapters, we explicitly discuss what understandings might be expected of students at age 11 years. As explained later when discussing the curriculum, there is in fact very little focus on uncertainty before 11 years of age. Indeed, the research on adult reasoning tends to show weaknesses that are similar to those for younger children, so in this particular area of knowledge it might well feel to teachers that they are starting from the beginning. As a result, we have no specific section on what children might already know and instead move immediately on to consideration about the difficulties that children (and adults) have with reasoning about uncertainty. We also note some of the ways in which learners' meanings for randomness mirror those of experts and offer useful starting points for teaching and curriculum development. The focus then changes to the way in which probability has been positioned in the curriculum as well as views about appropriate ways of thinking about progression in student development of reasoning about uncertainty. We then discuss various teaching approaches before the summary of the main issues in the chapter and some areas of study where additional evidence is needed.

Many of the studies reviewed in this chapter are also considered in the key reading by Bryant and Nunes (2013). We recommend this key reading to those who wish to consider more deeply the validity of this research, read further connections between the studies, and begin to delve into some areas such as correlation, not emphasised in this chapter.

## The nature of reasoning about uncertainty

The challenge of making sense of situations that are unpredictable pervades everyday living. One example is games playing. When playing chess, a player moves a chosen piece with complete information about the board and the rules. Chess is sometimes described as a total information game because you can see all the pieces on the board before choosing how to make your move. In contrast, many card games, such as whist or contract bridge, are formulated so that you only have incomplete information. For example, you do not know the complete deal and so have lack of knowledge about the cards that other players hold. This lack of information creates uncertainty, an additional dimension to the game. Sporting contests also possess uncertainty, which many people find compelling. A player who is able to reason about uncertainty will be able to adopt strategic approaches, which will improve the player's success rate in the long term, even if in the short term the optimum strategy might be 'unlucky'.

Uncertainty is also apparent in everyday decision making. Weather forecasters now regularly associate a probability with their specific prediction (as in Figure 7.1) and that may influence people in decisions about which activities to pursue that day. Doctors and surgeons offer rates of occurrence of side effects and complications when offering a range of actions that the patient might take

**Today's weather**

Cloudy, showers around

| Max 15°C (59°F) | | |
|---|---|---|
| **Probability of rain:** | 78% |
| **Maximum Temperature:** | 15°C (59°F) |
| **Wind force (Beaufort):** | 3 |
| **Wind description:** | Gentle Breeze |
| **Wind direction:** | SE |
| **Sunrise at:** | 5:05 am  ▲ Top of page |

**Figure 7.1** An illustration of how probabilities are used sometimes in weather forecasts. This is taken from 'Transport for London' on 18 March 2012 (http://www.tfl.gov.uk/tfl/weather/).

including medications and operations. In both the cases of the weather forecast and the medical situation, decisions might be made that consider both the likelihood of the various contingencies and the benefits or costs should they happen.

In games playing, the player may mathematise the situation, for example by using probabilities, whereas the weather forecaster and doctor are providing information already based on such a mathematisation. A player who is unable to create such a mathematisation, or a patient who is unable to understand one when offered, may be disadvantaged. In mathematics, a random variable is used to represent a situation that could occur in several different ways and each possible outcome would typically have a probability associated with it. The combination of possible outcomes and probabilities is referred to as a probability distribution. This mathematical notion of randomness can be used to model phenomena and this is what the weather forecaster and doctor do when they give likelihoods or rates.

Modelling with uncertainty is a mathematical task that informs much professional life, including the actuary who needs to calculate life expectancies and the financial advisor who considers investments in the stock exchange. In modern society, the mathematics of randomness is programmed into computers to run such models. In fact, the communication of uncertainty to computers when creating such models requires the mathematical notion of a random variable. This type of randomness is sometimes referred to as pseudo-random since it is argued that a machine must be able to compute the next number in the sequence formulaically. Nevertheless, from the point of view of the user or programmer, the next number in the sequence is as unpredictable as the next throw of a die or the uppermost face of the next toss of a coin.

In a sense, neither the throw of the die nor the toss of the coin are intrinsically random. A decision is made in the context of the event of tossing the coin or throwing the die to view the situation as suitable for the application of a random mathematical model. One can imagine, as a mind game, knowing everything about the throw of the die from the strength and direction of the throw to the coefficient of friction when the die hits the table so as to be able to compute an exact prediction of which face of the die will show uppermost. In such circumstances, it may be appropriate to use a deterministic model to analyse the problem. In this sense, randomness or determinism lies not in the situation itself but in the decision about which mathematical or scientific model to apply.

Thus, reasoning about uncertainty requires an appreciation of when it might be appropriate to apply a random model and when not. Often, a situation might not be recognised as being open to a random analysis. The games player might enjoy their game and see what happens as essentially a matter of luck. They will be unlikely to be successful in the long term but of course that might be of no concern to them. The notion that uncertain events, if repeated sufficiently often, become predictable, referred to as the Law of Large Numbers, is a key understanding in reasoning about uncertainty and underpins why strategic thinking when optimised will succeed in long term games playing. Of course, often events are not repeatable and so it becomes more a matter of faith that there is nevertheless some sense in deploying the optimum strategy.

## Students' reasoning about uncertainty

Since the mid-1970s, there has been a huge research effort to understand how students make sense of scenarios that could be modelled as chance situations. Much of this research has alerted educators to the fallibility of students, what they do *not* know, but more recently there has been increased emphasis on what intellectual resources students *are able* to draw upon. Much of this effort has been inspired by the seminal work of Piaget and Inhelder (1975) and that of Fischbein (1975).

An important aspect of the work by Piaget and Inhelder was based on understanding of random mixtures when balls were tilted inside a box. According to Piaget and Inhelder, 11-year-old students would normally be at a stage in their development where they would anticipate that, as the balls became progressively more mixed, they would be very unlikely to return to their original positions. By simulating the fall of raindrops on paving squares, Piaget and Inhelder inferred that students at this stage could construct irregular distributions that progressed towards regularity with increased raindrops. That is to say, the students would show paving squares with more or less equal numbers of raindrops when there were large numbers of raindrops. Piaget and Inhelder claimed that the students were basing their results on an intuitive sense of proportion without reflection on the random nature of the rainfall rather than an appreciation of the Law of Large Numbers where randomness would be an integral component when modelling the rainfall.

According to Piaget and Inhelder, some students will develop between ages 11 and 18 years to a stage of formal operations and these students will be able to coordinate knowledge about proportion, combinations, and randomness to construct probabilistic operational knowledge. This analysis positions probability as one of the most sophisticated achievements in mathematical knowledge development at school.

In contrast, Fischbein's work concluded that even pre-school children had some intuition of relative frequencies and concluded that the curriculum failed to nurture these early ideas by emphasising causality at the expense of the stochastic.[1] Fischbein's analysis, in contrast to that of Piaget and Inhelder, suggested that it might be possible to design teaching approaches that would support students in constructing normative intuitions. In this chapter, we begin by examining research that has built upon or critiqued the contrasting positions above and conclude with research on teaching approaches that might respond to the challenge of helping students to reason about uncertainty.

## Chance and fairness

People of all ages are constantly confronted with decision-making situations to which they might apply strategic thinking. Often the decision needs to be made when not all of the information is available. In games-playing, it is not difficult to understand the difference between a game such as chess, when all of the information is available setting a challenge to the logical mind, and many card, board, and video games where the lack of information can generate excitement. It may be less evident how to manage the latter type of games in a strategic way.

---

[1] The term stochastic has been used in different ways within statistics education research. Here, I am referring to any aspect of statistics or probability that draws upon randomness.

Yet coping with incomplete information is the typical situation in our everyday lives when social, health, and work-related decisions need to be made.

If Piaget and Inhelder were right, most people would be constantly confronted with decision-making situations that ideally would demand a probabilistic analysis prior to the development of the corresponding intellectual apparatus associated with the onset of formal operational thinking. Neither would the situation be resolved if Fischbein were right, as then schooling would not be providing students with the intuitive sub-strata that would enable them to respond in appropriate ways. So, the question that emerges is how do 11–18-year-old students, in these less than ideal circumstances, deal with situations that demand judgements of chance?

Kahneman *et al.* (1982) conducted a series of studies in the 1970s and 1980s, which perhaps threw light upon this question. They catalogued the heuristics (that is to say, intuitive rules-of-thumb) that people of all ages use when making judgements of chance. For example, they identified the *availability* heuristic, by which people make judgements of chance by evoking from long-term memory similar events. Another example is the *representativeness* heuristic, in which people expect the actual outcomes to represent closely the perceived sample space. Thus, in Roulette, one might expect the ball to land with regular occurrence on red and black numbers since red and black (ignoring the zero) make up the sample space.

www.nuffieldfoundation.org/uncertainty-4

It could be argued that such heuristics were developed as an intuitive means to guide decision-making in the absence of formal operational thinking. Kahneman and Tversky, in an extensive body of work, claimed that such heuristics were often effective in steering judgements of chance but that they contained systematic bias that would in some conditions lead to incorrect judgements. In using the availability heuristic, incidents will often be recalled because of their particular salience rather than frequency and so some bias in judging the likelihood may creep into the judgemental process. For example, when deciding whether flying is safe, a recent news-worthy aeroplane disaster might influence the decision in ways that do not relate to the likelihood of an accident. In using the representativeness heuristic, too much attention might be placed on the sample and too little on how outcomes might not represent the sample very closely because of randomness. For example, the representativeness heuristic might lead a gambler to bet rashly that, after a long run of black numbers, a red number would be more likely to occur in the belief that the outcomes should reflect the sample space which is made up equally of red and black. (This tendency to bet on the outcome that has not occurred is referred to by researchers as *negative recency*. The opposite strategy would be to bet on the same outcome occurring again and this would be referred to as *positive recency*. There is perhaps a better argument for positive recency since the string of similar results might point to a bias in the roulette wheel itself!)

Other researchers have identified people's fallibility in making judgements of chance. Piaget and Inhelder's developmental theory would suggest, at the stage of concrete operations, a natural proclivity or tendency to construct deterministic accounts of phenomena. Fischbein would argue that any such tendency is often reinforced in any case at school. Konold (1989) noted a tendency for people to focus on the determined outcome, ignoring or simply not recognising opportunities for a probabilistic approach to the situation. It worked and that's the end of the matter! In the outcome approach, a decision would be right or wrong according to whether it led to success rather than whether, in a strategic sense, it would be the decision that would lead to success more often if repeated many times. Since it is often the case that decision-making scenarios are one-offs, it may seem obscure to consider a strategic approach that envisages the repetition of an event that will not be repeated, but that is what strategic thinking often needs to do.

Lecoutre (1992) offered her subjects the opportunity to draw different coloured sweets from a bag – two were orange and one was lemon. They were asked which colour was more likely. Often responses demonstrated the equiprobability bias by which it was claimed that the odds for each colour were equal or that it was not possible to predict, as it was just a matter of chance.

Lecoutre then provided simple shapes that could be put together to form a house shape, jigsaw fashion (see Figure 7.2). She organised this task so that the combinatorial analysis for the shapes problem was equivalent to that which might be conducted in the sweets task. The proportion of ways of choosing pieces that could make a house was equal to the proportion of ways of choosing an orange sweet. She showed that in the two mathematically isomorphic situations people were more successful in solving the problem that depended only on combinatoric logic than in solving the one that depended on exactly the same combinatoric logic but set in a chance context. In the latter case,

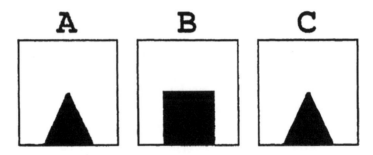

**Figure 7.2** The shapes used in Lecoutre's experiment (1992) allowed a house to be drawn if shapes A and B or B and C were used and a rhombus if A and C were used (p. 561).

the tendency for people to regard the answer as 'just a matter of luck, 50-50, it could be one or the other' resisted modification.

www.nuffieldfoundation.org/uncertainty-5

Studies by Green (1982) and Engel and Sedlmeier (2005) suggest that there is no improvement by age in students' understanding of randomness. Van Dooren *et al.* (2003) have argued that research in geometry and elsewhere on the over-generalisation by students of proportionality to situations which are not proportionate extends to the domain of probability. Furthermore, they argued that this linear illusion could explain many of the naïve conceptions discovered by Kahneman and Tversky and others. They believed that the emphasis on proportionality in schools contributes to this error, rather like Fischbein arguing that a curriculum focus on determinism undermines the development of stochastic thinking (but see Chapter 3 for the problems of establishing multiplicative reasoning in the first place).

If it is the case that new approaches to teaching could better nurture early intuitions, research needs to identify what students already know as this knowledge would, from a constructivist position, provide the starting point for pedagogic innovation. Pratt (2000) carried out clinical interviews of 10–11-year-olds and found evidence that these students had intuitions for randomness that were remarkably similar to those more enculturated into mathematical practice. According to this study, it would be reasonable to suppose that many students at age 11 would regard a phenomenon as random when it appeared to be unpredictable, irregular (in the sense that apparent patterns in results were not sustained), uncontrollable, or fair (usually based on a symmetric appearance). Pratt observed that these intuitions were local in nature in the sense that they appealed to the immediate. He also found that these students did not articulate *global* meanings for aggregated long-term behaviour.

These findings fit with a study of student's beliefs and judgements about fairness of dice (Watson and Moritz, 2003). In this study, school students above age 10 years, were found to judge fairness according to appearances of symmetry with experience being of secondary importance, perhaps limited to a few throws of the dice with no systematic recording of results.

Pratt responded to the apparent emphasis on local meanings by developing pedagogic approaches to nurture global meanings and these will be discussed later.

## Sensitivity to context

The development of new teaching approaches needs to recognise the sensitivity that students have to what might seem to experts to be superficial changes in the problem context. Research would seem to suggest though that learning what is superficial and what is part of the intrinsic mathematical structure is what the expert knows and what the naïve user does not necessarily know.

Konold *et al.* (1993) reported on the inconsistency with which students respond to test items so that small changes in the wording of the question resulted in very different responses, some apparently displaying the outcome approach in one form of the question but then seeming to make use of the representativeness heuristic after a small change in the wording. Later, Konold (1995) reported that students appear to hold simultaneously multiple and sometimes contradictory beliefs about a particular situation.

Similar results were found by Pratt (2000). In clinical interviews with 10–12 year olds, he found that the students would articulate many different meanings for randomness and that to him these meanings would at times appear contradictory. He went on to challenge the students by giving them different simulated random generators, referred to as *gadgets*, (a coin, spinner, and die, for example, as in Figure 7.3), designed so that the mathematical structures were presented in almost identical fashion and similar tools were provided in each case. The gadgets were incorporated into *ChanceMaker* (http://people.ioe.

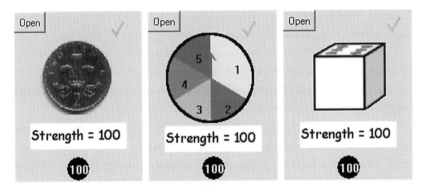

**Figure 7.3** Three of Pratt's (2000) gadgets, which simulate the behaviour of material coins, spinners, and dice.

ac.uk/dave%5Fpratt/Dave_Pratt/Software.html). As the students moved from one simulation to another, they would typically seem to ignore any progress made in the earlier problem contexts, acting as if the new problem context had no connection with earlier experience. Presumably the differences in appearance and animation of the simulations made more initial impact on the students' thinking than did the consistent mathematical structures and exploratory tools. In fact, gradually the students did begin to re-use meanings constructed from the earlier work, suggesting that under certain pedagogical conditions, it is possible to promote a more consistent use of normative meanings. We discuss the pedagogic implications later.

www.nuffieldfoundation.org/uncertainty-1

## Sample space and probability distribution

Piaget and Inhelder reported how the intellectual demands of constructing probabilistic knowledge included not only an understanding of randomness, discussed above, and proportion (possibly over-generalised) but also combinatorics. Knowledge of the latter enables the delineation of the sample space associated with a random variable. For example, whereas constructing the sample space for the throw of a single die might be trivial for older students, there might be rather more difficulty in identifying the 36 items in the sample space for the throw of two dice. Shaughnessy and Ciancetta (2002) investigated students' responses to a game in which two spinners had a 50% chance of pointing to either a black or a white sector. The students were asked whether there was a 50% chance of both spinners indicating black or not. Although 90% of those enrolled in upper level courses age 11 to 13 years gave correct answers, only 28% of ages 10–12 in the introductory level were correct in their response and a very small number (8%) gave good reasons. A large proportion of 7 to 12 year olds did not consider the sample space in giving their answers.

Batanero *et al.* (1997) identified typical errors in combinatorial reasoning as (i) a failure to manage correctly the order of elements, for example not including both (1, 2) and (2, 1) in the sample space for the throw of two dice, perhaps resulting in a sample space of 21 elements; and (ii) failure to work systematically, generating errors of repetition or omission.

 www.nuffieldfoundation.org/uncertainty-2

The apparent complexity of combinatoric thinking is not the only difficulty associated with sample space. Students need also to recognise the relationship between the sample space and the distribution of results. Even in the case of simple events, such as the throw of a single die, students can go wrong. In Pratt's study (Pratt and Noss, 2002), the students used computer-based simulations of dice and other phenomena (called *gadgets* as before). They were able to the edit the gadget's workings, which controlled how the gadget worked (see Figure 7.4). A normal die's working box might read 'choose-from [1 2 3 4 5 6]' but this could be edited to read any other configuration, such as choose-from [1 2 3 4 5 6 6 6], which could be understood as an eight-sided die or as a six-sided die with a bias towards 6s. The students did not unproblematically connect the sample space to the data distribution. For example, the lack of uniformity in the pie chart of results from small number of simulated throws of a single die led the students to edit incorrectly the sample space away from [1 2 3 4 5 6]. Later, when the students explored a biased die with a sample space of [1 2 3 4 5 6 6 6], they expected the distribution to be uniform for large number of simulated throws. We will discuss

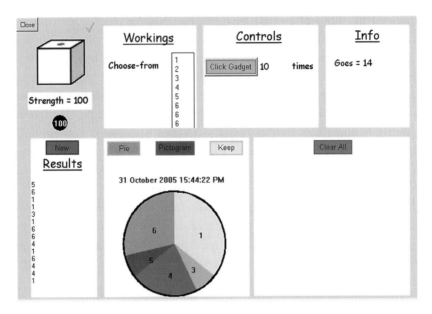

**Figure 7.4** The tools, including the workings box, inside the die gadget.

later how Pratt's students gradually developed normative ways of thinking about the relationship between the sample space and the data distribution.

For compound events the relationship between the sample space and the data distribution is further complicated. The outcome approach (Konold, 1989) and the equiprobability bias (Lecoutre, 1992) discussed earlier will have the effect that some students will claim that there is no reason to expect one outcome over any other. Fischbein *et al.* (1991) also substantiated this finding when they asked students to compare the likelihoods of a 5 and a 6 with two 6s when throwing two dice. Fischbein identified a more sophisticated, but nevertheless wrong, argument put by students that each compound outcome is equally likely because each simple outcome is equally likely and the tosses are independent.

The sample space and the associated probability distribution are similar but typically a sample space is expressed in terms of the equally likely outcomes whereas a probability distribution is usually represented by associating a probability with each unique outcome of the random variable. In the long-term, there is a sense in which the data distribution comes to match the probability distribution but it is perhaps less obvious how the sample space and the data distribution are related, except in the case of simple events. Even so, Prodromou and Pratt (2006) identified how 14-year-old students employed different approaches in order to make sense of the connections between what they termed the modelling and datacentric perspectives of distribution. Some students saw the data distribution as gradually morphing into the probability distribution with the latter being regarded as a target to which the data distribution was headed. They speculated that such a perspective focuses on the variation in data and positions the process as an emergent phenomenon. Other students saw the connection in the opposite direction, with the modelling distribution as an intention generating the data distribution. Such a perspective seemed to focus on randomness. Later, we discuss how they saw the role of causality in coordinating these perspectives.

 www.nuffieldfoundation.org/uncertainty-3

## Progression in reasoning about uncertainty

In this section, we discuss two types of research, the development of cognitive models of probabilistic reasoning and arguments about the curriculum.

### Cognitive models of probabilistic reasoning

There has been considerable research aimed at offering hierarchies in students' probabilistic reasoning, based on an analysis of success rates in different items that test knowledge of various probabilistic concepts. Many of the developments in this area have been influenced by the SOLO (Structure of Observed Learning Outcomes) taxonomy (Biggs and Collis, 1982). The SOLO taxonomy would predict that students are moving during ages 11–18 years from the concrete-symbolic to formal modes of functioning in line with Piagetian stages of development. Each of these modes, according to this approach, contains within it a cycle of five hierarchical levels: (i) *prestructural*, ongoing from the previous mode, in which students can be easily distracted by irrelevant aspects of the task; (ii) *unistructural*, in which the student focuses on one aspect of the task; (iii) *multistructural*, in which several aspects are recognised but not integrated; (iv) *relational*, when those different aspects are in fact integrated; (v) *extended abstract*, when the student generalises the structure and moves into the later mode.

These hierarchies demonstrate what students appear to find difficult and may seem to offer a progression that could help to structure curricula. Older students have already been influenced deeply by the curriculum and so it is difficult to separate out curricula and developmental influences. Nevertheless, the SOLO taxonomy provides a generic framework that can be applied to research and curriculum analysis.

Consider for example the students in Pratt's study (Pratt and Noss, 2002). They began by exploring contextual factors such as the strength with which the simulated spinner was thrown before gradually focusing on the number of trials and recognising what Pratt referred to as the N resource, expressed as 'the more times you throw the coin, the more even is its pie chart'. When Pratt referred to resources here, he imagined pieces of knowledge that were available for students to draw upon when trying to make sense of phenomena. Thus, N is a piece of knowledge, developed by the students through interaction with *ChanceMaker*, that could be seen as a situated version of the mathematical Law of Large Numbers.

www.nuffieldfoundation.org/uncertainty-1

By focusing on one aspect to the exclusion of irrelevant contextual factors, these students had, in SOLO terms, moved from the prestructural to the unistructural level. Later the students focused on the sample space of the spinner and constructed D, 'the more even are the spinner's sectors, the more even is the pie chart'. Pratt referred to D as a piece of knowledge that could be seen as a situated version of the mathematical notion of distribution.

At this point, the students were working with two distinct aspects of the spinner's behaviour; in SOLO terms they were operating at the multistructural level. A few students coordinated these ideas at the relational level to construct N.D, 'the more the more times you throw the spinner, the more even is the pie chart, provided the spinner's sectors are even'. In fact, students developed these ideas as they moved between the different random generators, typically reverting to prestructural ideas when first moving to a new generator. Students at some point recognised the relevance of N, D, or N.D as articulated in a previous generator.

The tendency to revert to prestructural ideas reduced once several gadgets had been experienced. Indeed, as Pratt's students began to recognise the broader power of N, D and N.D to explain behaviour across a range of generators, they were perhaps beginning to operate at an extended abstract level.

This research suggests that the cycles through prestructural, multistructural, and relational levels may not be completed before a new cycle begins. Nevertheless, the example illustrates how the SOLO taxonomy can be applied to research that never sought to identify a progression in student's understanding. More importantly, it shows how progressions in learning might be identified and used to inform teaching and the construction of curricula.

## Curriculum content

In the light of the seminal research by Piaget and Inhelder, curriculum developers regarded probability as inaccessible to children up to age 11 years. It could be argued that this was a misinterpretation insofar as Piagetian theory asserts that a learner must pass through pre-ordained stages of development but does not pin those stages to particular ages. Nevertheless, it is difficult to align findings that probability is a late development with teaching the subject at a young age.

Perhaps persuaded by Fischbein's findings that schools might support very young children's intuitions for relative frequency, criticism of the status quo mounted and elementary probability was incorporated into the curriculum in the UK and other countries. There was though a perception in the UK that children continued to find probabilistic ideas inaccessible, and in the face of pressure from competing topics, gradually probability has been removed from the curriculum in the UK to the extent now that there remains very little emphasis in the National Curriculum at Key Stages 1 and 2 (up to age 11).

It remains unclear whether this experience supports the Piagetian view since it could be argued that the form of the experience of probability inserted temporarily into the curriculum for children below age 11 did not match with the aspirations of Fischbein when he advocated the introduction of chance experiments. Certainly, Threlfall (2004) argues that the type of experience offered to young children was not likely to result in improved knowledge about the mathematical notion of probability. However, Threlfall goes on to suggest that the simple probability tasks *cannot* challenge everyday notions and therefore concludes that perhaps probability is best left until later schooling when students might engage with more complex tasks.

As expertise in the use of technology to provide novel learning experiences increases, perhaps it is best to remain open-minded about the possibility of a return of probability to the curriculum for children below age 11 but the current position is that teachers can expect students age 11 years to have little formal experience of probabilistic ideas, although they will have everyday experiences that will shape how they interact with learning opportunities offered in the curriculum for 11–18 year olds.

At this later level, the National Curriculum in the UK has focused on:

- relative frequency as an estimate of probability;
- the use of subjective estimations;
- the addition of probabilities for mutually exclusive events;
- the multiplication of probabilities of independent events;
- the use of tabulation or tree diagrams for managing compound events.

Compared to some other curricula, such as those in Australia and the USA, there is less emphasis in the UK on modelling and the social uses of chance (see Table 20.2, p. 914 in Jones *et al.*, 2007).

The danger inherent in the UK curriculum is that probability teaching focuses on the algebra of probability, which is meaningless if probability is not seen as useful for making judgements in real life and for solving problems. Such a danger might manifest itself in a tendency to apply in rote fashion the laws of probability in order to make computations, especially when probability is perhaps seen as difficult, even counter-intuitive, and when teachers are under pressure to improve examination success rates. Such procedural approaches have been reported by several researchers. For example, Greer and Mukhhopadhyay (2005) observed the 'exposition and routine application of a set of formulas to stereotyped problems' (p. 314). Probability is increasingly isolated in the UK curriculum to the extent now that it might seem from the student's perspective as an esoteric activity focused around arcane objects like coins and dice.

In contrast, when presented as a modelling tool, probability can be linked to statistics as when modelling empirical data or making inferences. For example, Konold is developing TinkerPlots™ (www.keypress.com/x5715.xml) to enable not only the graphing of data but also its modelling with uncertainty. TinkerPlots™ is used extensively in Chapter 6 to illustrate exploratory data analysis (Chapter 6, Figures 6.1–6.6). Version 2.0 allows the user to create models of phenomena through building probability distributions based on urns, spinners, or histograms. This version is used in Chapter 6 to depict re-sampling (Chapter 6, Figure 6.7). The model created in Version 2.0 can then be executed

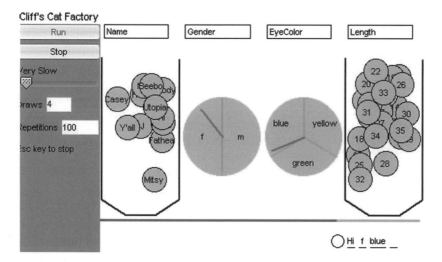

**Figure 7.5** A machine designed in a beta version, later published as TinkerPlots™ Version 2.0 for making cats with various attributes.

repeatedly to create data distributions. In one paper (Konold *et al.*, 2007) for example, young students are encouraged to design machines that make cats with various attributes (name, gender, eye-colour, and length or whatever attributes might be chosen by the students). Figure 7.5 depicts one such machine.

www.nuffieldfoundation.org/uncertainty-5

www.nuffieldfoundation.org/uncertainty-7

http://www.keycurriculum.com/products/tinkerplots

Another example of using probability to model phenomena (Prodromou and Pratt, 2006) allows students to model the actions of basketball player (Figure 7.6). The user can control the aspects of the throwing action (speed, direction, distance from basket) to try to complete a successful basket. In one mode, the result is not entirely determined by the throwing action and so is subject to other unknown factors. The students explore this stochastic situation through control over the spread of the distribution of aspects of the throwing action. In these examples, probability becomes a modelling tool when not everything is known about the phenomenon in question. (The Basketball software is freely available from http://people.ioe.ac.uk/dave%5Fpratt/Dave_Pratt/Software.html.)

www.nuffieldfoundation.org/uncertainty-1

The tendency for probability to become increasingly disconnected from the rest of the mathematics curriculum has perhaps been exaggerated by the exploratory data analysis movement, which has sought to avoid the need for probability, as discussed in Chapter 6. The curriculum for 11–18 year olds in its current form arguably does not support the teacher who wants to make probability relevant and meaningful, connected to statistics as a tool for modelling real-world phenomena.

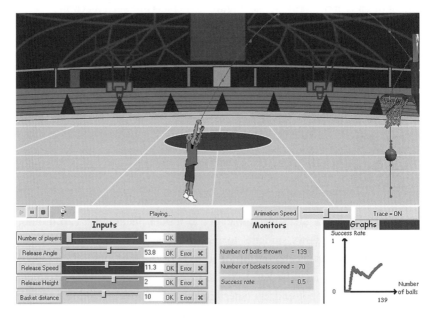

**Figure 7.6** In Prodromou and Pratt's Basketball simulation (2006), the release angle, speed, height, and distance can all be varied using the sliders. When the error buttons are pressed, random variation is introduced into the flight of the basketball.

# Teaching approaches to reasoning about uncertainty

The following pedagogic heuristics are reported in the research literature as especially felicitous in helping students to develop probabilistic intuitions.

## Working with naïve conceptions

The constructivist movement, informed by the work of Piaget, led to the notion that students' ideas must be challenged in order to promote accommodation through reflective abstraction. The psychological literature on probabilistic understanding is in fact replete with catalogues of human fallibility, reported as misconceptions, some of which were discussed earlier in this chapter. Often the education literature advocated the removal or overcoming of these misconceptions. Smith *et al.* (1993) argued persuasively that such a conclusion was in fact anti-constructivist since it posited no basis from which new ideas might be developed. As far back as 1998, Castro (1998) reported that a conceptual change approach to teaching improved intuitive reasoning about probability when compared to more conventional approaches. Castro emphasised the need to generate cognitive conflict and build on students' ideas.

Researchers such as Konold (1991), Pratt (2000) and Abrahamson and Wilensky (2007) have sought to build on naïve conceptions, regarding them as potential sources of learning, rather than as obstacles to be removed. For example, young students may have expert-like perceptions of randomness as unpredictable, uncontrollable, and irregular, as proposed by Pratt (2000), but may not recognise the limitations of these perceptions to the short-term. Similarly they may not recognise that even unfair situations might be random. When such naïve conceptions are over-generalised, the pedagogic challenge is to introduce the student to the limitations of the idea, rather than to seek to eradicate the conception.

On the other hand, students may not appreciate the power of their idea. It is normal for a student's new ideas to be situated in those contexts in which they were learned. For the student, it is perhaps surprising when the proportion of sixes on the die equals the proportion of each of the other numbers, but for him or her this may be a generalisation about dice. Yet truly powerful mathematical ideas will often transgress that context, possibly to the extent that the mathematician sees the idea as de-contextualised. The pedagogic challenge is then to seek to introduce the student to the power of the naïve conception, so that the result about the die has its equivalent result for coins, spinners, and many other contexts, until eventually it can be seen by the student as general, the Law of Large Numbers.

Not all naïve conceptions are equally useful. Some may simply be unproductive. The teacher must be alert to ideas that have potential for what Prediger (2008) calls horizontal development and to use interventions designed to support that development. Such horizontal development will not remove the unproductive naïve conceptions, though they may become less powerful for the student in explaining stochastic phenomena.

 www.nuffieldfoundation.org/uncertainty-6

## Broadening the domain of applicability

Meanings for what might be seen as the same concept by the statistician are often attached in the learning process to the situation. Pratt and Noss (2002) reported how meanings abstracted from working with one virtual random generator were not easily transferred to a different device. For example, a learner might know that when a fair die is thrown many times the proportions of the scores become roughly equal but might not yet recognise that s/he could apply that idea also to say a spinner with equal size sectors. Learning to re-use meanings across situations, according to Pratt, requires careful pedagogic support, by allowing students space and time to explore tasks that have the same underlying mathematical structure but which are set in distinct situations.

Another aspect of this difficulty of seeing that apparently different situations can be approached through the same probabilistic analysis is inherent in Konold's Outcome Approach (1989). Here, students may not recognise that the problem is amenable to a stochastic analysis. Fast (1999) advocates the use of analogies to bridge between situations that are more evidently stochastic, such as when they refer to lotteries, dice, spinners, and so on, with situations that are less explicitly random, which actually embrace most real-world applications of the probability. First, Fast showed how students were much more successful on problems that were explicitly probabilistic. Later, Fast showed that when the teacher exposed the connections, performance on analogous tasks also improved.

It appears that both Pratt's approach and that of Fast provide support for the broadening of what Pratt and Noss (2002) call the *contextual neighbourhood* associated with the meaning, through sequences of somehow similar tasks.

## Using simulations

Research using calculator-based and computer-based simulations offers increasing evidence that teachers might adopt such tools to support student's probabilistic understanding.

Tools, such as graphical calculators, Probability Explorer (www.probexplorer. com/), Fathom™ (www.keypress.com/x17675.xml), and TinkerPlots™ (www. keypress.com/x5715.xml), are generic insofar as the student might use them to build their own simulation of any problem for which a stochastic model could apply. Ongoing work in this area has led to the formulation of a teaching sequence in which first the real-world problem is introduced, second a simulated model is built, third the model is used to generate data to allow inferences, fourth the validity of the conclusions are re-considered given assumptions in the model, and only then, as a fifth step, is the problem analysed formally using probability and statistical methods (Batanero *et al.*, 2005). In other words, experience with the simulation provides initial intuitive substrata to act as a resource to provide meaning in the formal analysis.

http://www.keypress.com/
x5715.xml

http://www.keypress.com/
x17675.xml

http://www.probexplorer.
com/

ChanceMaker (http://people.ioe.ac.uk/dave_pratt), designed originally for research purposes, is not generic as it allows only exploration of a limited and fixed number of situations. Nevertheless, the approach requires some simulation building since the students are required to fix and explore the sample space (called the *workings box*) of the various gadgets that simulate real-world random generators. Nevertheless, the first four steps of the above instructional sequence seem to apply to the envisaged use of ChanceMaker and could perhaps prepare for more formal work on probability.

 www.nuffieldfoundation.org/uncertainty-1

A common aspect of all of these tools for simulation is that they give the student access to repeated trials. Teaching strategies in which results from around the class are merged to create larger datasets certainly have their use but often the number of trials needed to be confident that long-term effects can be perceived is very large. Technology provides a means for conducting such large-scale experiments in acceptable time scales.

Additionally, the power of technology can be harnessed to generate many different graphical and numerical representations in short time periods so that the visual feedback can help students to recognise weaknesses in their current understanding.

Finally, technology supports exploration of the sample space. Physical objects are usually not amenable to investigation that involves changing the configuration of the sample space. The malleability of virtual resources can facilitate such experimentation, which might be vital in coming to realise the connection between the sample space and the resulting distribution of results. This is discussed further below.

## Summary

In trying to make sense of randomness, young students will draw on whatever mental resources they have and these are inevitably based on the unsystematic *ad hoc* experience that makes up all of our everyday lives. Such experiences do not necessarily generate normative ways of thinking about the stochastic, resulting in the heuristics that are inherently biased, and an unpreparedness to recognise situations as amenable to a probabilistic interpretation.

Schooling can provide a more scientific approach by intervening in that experience. As evident in its title, there has been an emphasis in this chapter on reasoning about uncertainty and what has emerged advocates the use of probability as a mathematical tool for modelling. Seen in this light, students can be introduced to the panoply of situations that extend well beyond the arcane world of coins, dice, and spinners. Nevertheless, these familiar contexts can provide a pedagogic resource for recognising their mathematical similarity and their connection to other situations where the randomness element may be more obscure.

Through a modelling approach, probability can be seen as more connected to the rest of the curriculum. Exploratory Data Analysis, as reported in the previous chapter, enables students to exploit new tools for making sense of data. Statistical work is invariably focused on inference, the power to make statements about a wider population on the basis of a sample of data. The role of probability here is to provide a reality check on whether the apparent trends might reasonably have been generated by chance.

In most situations drawn from the everyday world of students, probability must be seen as a subjective measure of likelihood and in this sense probability is a tool for modelling uncertain situations. (Betting and gaming is the exception in that it offers a meaningful context in which probability might not be seen as subjective.) Adolescents especially are often curious to explore moral and ethical issues and to model such areas mathematically often demands the use of probability. A clear case is in the use of risk. Although risk is not a well-defined concept, it certainly involves chance and harm, and risk-based decision-making can provide a motivational context for subjective probability.

 www.nuffieldfoundation.org/uncertainty-8

 www.nuffieldfoundation.org/uncertainty-9

The teaching of probability appears currently to be limited to the algebra of the laws of probability and to the procedural use of representations such as tree diagrams. To escape these limitations and to connect probability to the rest of the mathematics curriculum, it needs to be seen as purposeful by students.

## Where additional evidence is needed

Many aspects of probability have not been thoroughly researched and we encourage teachers to build on the above ideas to conduct action research into how their students appear to respond in novel situations.

- Much of the work on heuristics by Kahneman and Tversky was conducted in order to identify how people make judgements of chance in short spaces of time with few supporting tools. Teachers are interested in how to change people's ideas. How do students' heuristics for chance change when provided with the opportunity to work with tools and with each other, supported by a teacher? ChanceMaker can be downloaded freely (http://people.ioe.ac.uk/dave_pratt). Pratt developed the tool working with 10–11-year-olds. What impact might working with ChanceMaker have on 8–10 and 12–14-year-old students?
- There is only limited evidence on how students think about situations that are partly determined. Statisticians model such situations through signal and noise or main effect and error. Whereas much of the variation might be attributable to a particular cause, nevertheless some of the variation is not accountable in that way and may be regarded as error. There has been little research of how students think about phenomena that are in a sense partially determined. TinkerPlots 2 ™ might be a suitable tool to explore such thinking.

 http://www.riskatioe.org

 http://understanding-uncertainty.org/

- There is little research on probability as a subjective measure. Risk-based decision making is one suitable context to explore. Research on the public understanding of risk is just beginning to emerge (see, for example, www.riskatioe.org and http://understandinguncertainty.org/). There is a need for more evidence about 14–18-year-old students' understanding of risk as well as suitable teaching approaches.

## Key readings

Bryant, P. and Nunes, T. (2013). *Children's understanding of probability: A report to the Nuffield Foundation*. Oxford: Oxford University Press.

The authors provide a review of the research literature on probabilistic reasoning, critiquing the findings and methods of much of the research referred to in this chapter and beyond.

Fischbein, E. (1975). *The intuitive sources of probabilistic thinking in children*. Dordrecht, The Netherlands: Reidel Publishing Company.

This seminal work provides the foundation for more recent studies that explore how pedagogic approaches might support the development of more sophisticated intuitions for probability.

Jones, G., Langrall, C. W., and Mooney, E. S. (2007). Research in probability: Responding to classroom realities. In L. F. K. Jester, Jr. (Ed.), *Second handbook of research on mathematics teaching and learning* (pp. 909–956). Charlotte, NC: Information Age Publishing.

This chapter provides a broad overview of research on the teaching and learning of probability.

Kahneman, D., Slovic, P., and Tversky, A. (1982). *Judgement under uncertainty: Heuristics and biases.* Cambridge: Cambridge University Press.

This is perhaps the best single source for the highly influential and controversial body of research on how people make judgements under uncertainty.

CHAPTER 8

# Functional relations between variables

## Introduction

This chapter continues the algebraic thinking that was described in Chapter 2, and also connects to some aspects of graphing described in Chapter 6. The central focus is on functions, including equations and graphs as representations of functions, and how these interconnect. We limit the content to what is normally in school mathematics, and draw on what research tells us about students' understanding and typical conceptual pitfalls.

### The nature of equations

Equations are statements of quantitative equality between two expressions. This equality can be dependent on particular values of the variables, such as in $2x - 5 = 11$, in which case we 'solve' to find the values, or can be a permanent definition or equivalence of two expressions involving several variables, in which case we can use one expression to evaluate the other. It is usual to call the latter kind 'formulae', as in $A = \frac{1}{2}bh$ for triangles, or $v^2 - u^2 = 2\ as$ and so on. However, in school mathematics the word 'equation' is often for expressions such as $y = 2x + 5$ in which the expression on the right is used to evaluate the dependent variable $y$, as a formula. An equation like $3x - 5 = 9 - 2x$ can therefore be seen as the intersection of two functions and solved graphically for $x$. There are therefore several connections between equations, functions, and graphs and the ways in which they are described can either aid or confuse understanding.

## The nature of graphs

Equations and functions express relations between quantities (and, in advanced mathematics, other mathematical objects) and graphs are one of the ways in which those met at school can be represented. The graphs relevant for this chapter are mainly those that relate two numerical variables, so that the relation is represented by a line or collection of points in two-dimensional space. The shape and connectedness of the graph depict how the variables relate to each other, how one changes as the other changes.

## The nature of functions

Functions are those relations in which the value of a variable is dependent on one or more other variables. Particular values for the independent variables generate one and only one outcome: the value of the dependent variable. More formally, a function is a mathematical relation such that each element of a given set (the domain) is associated with an element of another set (the co-domain). The subset of the co-domain onto which the domain is mapped is the range. In school mathematics this very abstract notion is 'a long time coming', so it is useful to think about the more informal meanings that pave the way (Elia *et al.*, 2007; Hitt, 1998; Leinhardt *et al.*, 1990; Slavit, 1997).

At various stages in the curriculum the roots of functions can be seen in:

- One-to-one or many-to-one mappings between sets, for example a mapping of six cakes to two children.

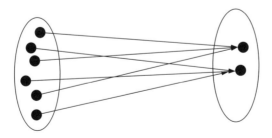

**Figure 8.1**

- Input/output machines with algebraic workings (e.g. $7(x + 5) =$ ).

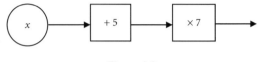

**Figure 8.2**

- Input/output 'black boxes' such as trigonometric or exponential functions.

We use the term 'black box' functions for those that are not calculable using school-level operations, that is those for which mathematicians rely on calculators, computers, and tables of values as appropriate.

- Expressions to calculate $y$-values from given $x$-values.

For example $y = 4x + 7$ being a formula for getting $y$-values from given $x$-values.

- Relations between particular $x$-values and $y$-values.

For example

| X | Y |
|---|---|
| 0 | 2 |
| 1 | 3 |
| 2 | 6 |
| 3 | 11 |
| 4 | 18 |

- Relations between a domain of $x$-values and a range of function values.

For example $f(x) = 0$ when $x$ is rational, $f(x) = x$ when $x$ is irrational.

- Representations of relations between variables in 'realistic' situations.

For example height of water in a cylinder[1] ($h$) = flow rate ($r$) x time ($t$) ÷ cross-sectional area ($a$) all in suitable units:

$$h = \frac{r}{a}t.$$

---

[1] It would be standard notation to use $h(t)$ to indicate that $h$ is a function of $t$, and this notation is potentially confusing as it could be interpreted by young students to mean height x time.

- Graphs which depict particular values.

For example:

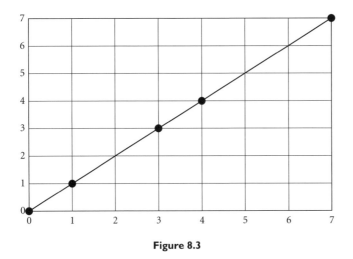

**Figure 8.3**

- Graphs which have particular characteristics.

For example a cubic has two turning points (which might coincide), and is continuous, and can intersect the x-axis in one or three places (which might coincide in adjacent pairs or all three).

- Graphs which can be transformed by scaling, translating, and so on.

For example sinusoidal functions are those obtained from scaling and translating (in the *x*-direction) the function $f(x) = \sin x$.

- Structures of variables defined by parameters and relations.

For example the parameters *a*, *b* and *c* define the function $f(x) = ax^2 + bx + c$.
Each of these views is limited to a particular way of seeing functions.

 www.nuffieldfoundation.org/functions-1

 www.nuffieldfoundation.org/functions-2

In general, there are shifts between seeing functions as instructions to calculate one numerical set from another, to seeing them as correspondences between sets, to seeing them as describing variations, to becoming familiar with functions (for which we need manipulable technology), to using functions to describe situations, and also to understanding functions (usually in graphical form) as objects in their own right (Tall, 1992).

## What do younger school students learn?

### Equations

In early years of school, there are ideas that could eventually contribute to learning about equations and functions. Students will have found unknown numbers in number sentences by using arithmetical facts and their inverses. They may have generated possible pairs for variables in expressions like $x + y = 6$, or $x - y = 4$, probably without using letters. Using a formula to generate values has probably been limited to calculating areas of simple shapes, conversions of units, budgeting, and making number patterns.

### Graphs

Early experience of graphs is likely to be limited to pictorial representations and representations of single variables such as in bar charts and pie charts. They may have met trend lines, or lines that map variables against time using experimental data. These may have been discussed qualitatively in terms of steepness, relative heights of data points, and special values, such as zero. They may know about the need to scale the vertical axis to fit the data, but not necessarily about scaling a horizontal axis. Apart from time, they are unlikely to have used a continuous variable horizontally, unless they have met this in science or had extensive graphical experience. They may also have met numbers on the horizontal axis treated as discrete variables for drawing bar charts. They may have met scatterplots, but probably not graphs for which the $x$-axis can be considered to represent an independent variable and the $y$-axis a dependent variable, unless conversion graphs are seen this way.

### Functions

They may have worked on term-to-term rules for simple sequences. They may be able to express functional relations in number sequences in words. They will have experienced one-to-one, many-to-one, and one-to-many mappings in contexts. They may have some sense of qualitative relations between variables in realistic phenomena, such as 'this goes up when that goes up'.

## Typical problems in reasoning about functions

Learning about functions, and learning to use them as modelling tools, is a long process which needs coordination throughout school and beyond. Students need to coordinate algebraic, graphical and numerical data in complex ways, and there are plenty of limitations which can become obstacles if they are not recognised and if there is not a consistent teaching approach taken throughout school. However, there is no research to suggest that one way of ordering the curriculum is any better than another (Leinhardt *et al.*, 1990). Whatever the order, the same limitations of prior knowledge need to be overcome at some stage and different approaches help students develop different conjunctions of understanding. The research tells us a lot about difficulties, but very little about successful ways to overcome them. Indeed, it tells us very little about how functions have been and can be taught, except in educational environments which are enhanced with special software, hands-on technology such as graphplotters and spreadsheets.

### Equations and formulae

Equations that can be solved using number facts do not easily develop beyond situations involving positive integers (Filloy and Rojano, 1989). Nor can they easily be seen to represent relations. This elementary number-based understanding of equations has to extend in two directions: towards algebraic manipulations that reveal unobvious values for equality, such as linear equations with negative and/or fractional solutions (Stacey and MacGregor, 1999), and towards being seen as instances of relations, such as knowing that $y + x = 5$ defines a relation between $x$ and $y$.

The pre-ICT approach to graphing functions used $y = f(x)$ equations as formulae to generate specific values for $y$. The data pairs $(x,y)$ were then plotted, and some appropriate interpolation and extrapolation curve was sketched. Understanding that every point on the ideal curve represents an instance of the original equation is hard to learn with this approach, because students tend to see 'points' as those specifically plotted and the 'lines' merely join the points.

The given domain, such as 'use $x = -10$ to $x = +10$', can also limit their perception. Using the '$y =$' equation of the function, when there is one, to generate data points focuses the mind on the values and calculations rather than the parameters. A smooth curve or straight line is usually expected, and used to check the calculations. This approach generates expectations that all functions have smooth recognisable graphical representations, and it also embeds an unhelpful notion that the data points are merely the framing for the graph, rather than particular instances of the function. The shift to seeing *all* points on a curve as instances, if this view can be achieved, provides the basis for using graphical methods to solve equations. A further shift is to seeing that a curve can extend outside the current visual frame, such as beyond the graph paper or beyond the screen. These extended perceptions can enable connections between the 'formula' meaning for the equation of a function and the 'finding the unknown' meaning for equations which follows from early school knowledge.

Ronda (2009) identified a sequence of 'growth points' in students' understanding of linear functions which can apply to functions more generally. Students have to understand the relation between coordinates for individual points, the fact that the same relation applies to sets of points, and that the equation and line represent this relationship and all the points that have it. They have to understand that there may be a procedure for generating the $y$-values, given the $x$-value, but they have to change from generating points to understanding the pattern between points, and then to working with the invariant properties of the whole relationship. Thus linear functions are a topic for which confusing the variables and parameters in an algebraic expression could cause problems (Sfard and Linchevski, 1994). For example, the roles of $y$, $x$, $m$, and $c$ in $y = mx + c$ are very different from each other, and as parameters the roles and appearance of $m$ and $c$ are different to each other. A related shift of understanding is from seeing equations as procedures to be 'done', to seeing them as representing relations, describing properties of relations, and then as functions. Clearly, a student who only has an arithmetical understanding of equations and a collection of remembered techniques, however fluent, is at a disadvantage since the fundamental understanding required is of relations between two variables.

## Interpretation of graphs

Students also have to adapt their experimental and statistical understanding of graphs as 'pictures' or measures of single variables, such as in bar charts and pictograms, to understanding that the $x$-axis can also represent a numerical variable, and that changes in one variable and associated changes in another can be represented by lines or curves (see Chapter 6 for more about statistical graphing). The 'single variable' assumption developed for bar charts persists (Bell *et al.*, 1987). Mevarech and Kramarsky (1997) asked if 13–14-year-old students could construct graphs of everyday situations. The context was to graph: 'the more I study the better my grades'. The students tended to draw one point to represent such statements, rather than a line, and this persisted in similar situations even after being taught about graphing variables.

'Time' on the $x$-axis is an intuitive and obvious idea, since the left to right movement represents chronological progression rather than being seen as a variable. This conception has to be upset to understand the abstract use of the $x$-variable (Janvier 1981). Interpretations of graphs as pictures is also common (Clement, 1985) and is emphasised by the one-variable approach, which leads students to believe that 'up' pictorially always means 'up' in a physical quantity. A related confusion, sometimes seen in bottle-filling problems, is that a high value must indicate a steep rate of increase or a steep slope always indicates a high value (Janvier, 1987). This is not a direct perception from the graph, but may be a linguistic confusion between 'faster' and 'higher', or even a physical sense of how the height of the graph is attained (Clement, 1985). Another persistent interpretation is that a travel graph has to look like the journey it represents (Gomez *et al.*, 1999), so the use of travel graphs to introduce graphing conventions does not necessarily help interpretation. Similarly, describing a graph as a 'path', such as when working with the idea of loci, can imbue it with irrelevant chronological meaning so that students think that graphs somehow start on the left and develop through time towards the right, and also look like a 'walk' along a path of a particular shape (Font *et al.*, 2010). Of course the graph of a circle is a circle, but the graph of distance travelled while going round a racing circuit is not the shape of that circuit.

To 'read' graphs of functions students have to understand correlation, either of abstract values or of 'real' data. They have to grasp that a graph is a visual representation of the relationship between a succession of points, but it needs interpretation. They then have to reason about points they can only imagine, between the existing points and also beyond them (Ainley *et al.*, 2000).

The $y$-intercept is the function value when $x$ is zero, and hence is the constant term in a polynomial function. Interpreting it this way has benefits for understanding families of functions that are related by translation parallel to the $y$-axis. Students often want the $x$-intercept to appear in a similar way to the *y*-intercept. For linear functions there is the implicit form: $ax + by = c$ in which $x$ and $y$ intercepts do act in similar ways, but for other polynomials explicit exploration of $x$-intercepts has to be recognised as more complex, along with the relation between solving equations and finding roots. Moschkovich (1998) argues that 'extending' meaning needs a planned approach. Her research showed 14 to 16-year-olds, who had learnt a lot about linear and quadratic functions, making incorrect simplistic assumptions about roots, intercepts and parameters. Students were asked to transform $y = x$ to match some given criteria or graph, and conflict was provoked by asking questions like 'Alfie says that $y = x + 5$ will go through (5,0), is he right?' One source of confusion is that translating a linear graph parallel to the y-axis affects the *y*-intercept, but the same visual appearance can be achieved by moving the graph parallel to the *x*-axis. Aspects of the graph that appear to have similar status have very different algebraic status (*x* and y intercepts) and objects having similar algebraic status, such as slope and *y*-intercept which both appear as parameters, have very different graphical status (Moschkovich, 1998, p. 194). Translating between algebraic and graphical systems therefore has hidden dangers and without this understanding the teacher can be unaware of what students are thinking. Greeno (1994) enabled 12 to 15-year-old students to understand the parameters by devising a gear system which physically enacted $c$ as displacements and $m$ as multiples of the independent variable so students could see that the relation is dependent on the parameters.

The study of quadratic functions could potentially move students beyond the confusions of the role of parameters in linear representations. First, translating quadratics is useful in showing the effects of different translation directions on position and intercepts, easily done with software (Zazkis *et al.*, 2003). Second, and often glossed over in textbooks, the different forms of the quadratic give different information in terms of their parameters:

- the y-intercept is visible clearly as $c$ in this form: $y = ax^2 + bx + c$ but the roles of $a$ and $b$ are obscure;
- the x-intercepts are visible as roots $a$ and $b$ in the factorised form $y = k(x - a)(x - b)$ but the y-intercept is not obvious; changing $k$ is interesting, particularly with software;
- the turning point (probably the most obvious visual feature of quadratics) is most visible as $(a,b)$ in the completed square form: $y = k(x - a)^2 + b$. The displacement of $+ a$ in the x-direction appears in $(x - a)$ which students find counterintuitive (Zazkis *et al.*, 2003).

We have deliberately used *a* and *b* in all three forms to illustrate how the choice of letters can confuse students, and that understanding mathematical text is about being skilful in knowing when the same letter means the same thing and when it means something different. The study of quadratics, therefore, provides an arena for students to learn to go beyond visual assumptions and to use parameters as clues or tools to interrogate and display the characteristics of a function.

 www.nuffieldfoundation.
org/functions-4

 www.nuffieldfoundation.org/
functions-5

## Drawing graphs

Drawing graphs from any data, whether from formulae or experiments, creates difficulties. Scaling and labelling axes appropriately is non-trivial for many students as it is related to their understandings of measurement and the real number-line. It is common for students not to relate axes to number-lines at all and mark them up with the relevant values for the current task spread out in various discrete ways along the lines. Some confuse a continuous number-line (a measuring scale) with a discrete set of number labels and draw axes as if for a bar chart, with a mixture of 0 and 1 as starting points (see Figure 8.4).

**Figure 8.4**

Once the number-line relation is established, students tend to assume a scale must be either in units or tens because of their knowledge of measurement (Williams and Ryan, 2000). Within pure mathematics the origin of the axes is always zero, but in plotting real data the starting point may not be an obvious 'zero', it could be 'Monday' or '20 cms' or '1967'. Often the *y*-intercept is the 'start' for a 'realistic' graph, yet algebraic graphs do not have left-to-right starting points (Davis, 2007). On the other hand, having situations in which the 'start' is not zero can break the common assumption that all graphs go through zero.

Experiences with realistic and algebraic graphing can contradict or reinforce each other, and students need to be able to work out when using knowledge from one is helpful and when it is not. For example, when plotting real data all you know are the readings, and they are probably approximate; what happens between these points has to be conjectured and the best you might get is a trend line. When plotting points from a formula, so the graph represents a mathematical relation, the points are exact and inter/extrapolation generates other data points precisely. A further confusion between graphs of realistic situations and those representing abstract functions is the inclusion of negatives which may have no meaning in real situations. Once these hurdles are overcome they have to be explicitly re-examined for non-polynomial functions, such as $|x|$ and $x^{(1/3)}$.

Students need to think about the relative scaling of axes and graphing software can obscure this by providing axes implicitly. On the other hand, use of such software can enable students to explore the effects of scaling – it depends on the task and the teaching. Some published resources reinforce the assumption that a line at $45°$ to the axes and through zero must represent $y = x$. This can lead to the assumption that reflecting in $y = x$ must result in visual symmetry, rather than relational symmetry; $x = y^2$ will only look shapewise like $y = x^2$ if the axes are identically scaled. A more frequent and critical confusion comes when calculating gradients using the actual distances on the graph rather than the distances represented by the scaled axes (Zaslavsky *et al.*, 2002).

The above comments show that graphing difficulties that appear to be technical and notational might actually be conceptual; students often need to apply an extended version of some earlier concept such as the number-line, the negative domain, relations between two variables, and scaling.

## Relations and functions

In early stages of learning about functions they can be seen either as a mechanism for pointwise correspondence of one variable to another (as an operation, function machine etc.) or as covariation of variables. The latter approach leads more obviously to focusing on increments and slope, but is harder to understand because the focus is on change rather than quantity (Mevarech and Kramarsky, 1997). Seeing functions only as mappings between sets of discrete data can reinforce a view that a function is a set of discrete data points. It can also reinforce a view that anything that can be mapped is a function (Spyrou and Zagorianakos, 2010).

Not all relations that can be mapped are functions, for example square root can only be a function if the codomain is restricted to either negatives or non-negatives. Some writers think that confusion between relations in general and the particular properties of functions is not worth worrying about (O'Callaghan,

1998; Confrey and Smith, 1995) and indeed the idea that $x^2 + y^2 = r^2$ is not a function unless the domain and codomain are limited is probably not as important as understanding that the equation describes a relation between $x$ and $y$, can be used to generate $y$-values from $x$-values and *vice versa*, and has certain properties, symmetries, and zeroes.

More critical is the erroneous idea that all functions are continuous, smooth, and calculable (Leinhardt *et al.*, 1990). Several realistic situations that can be expressed as functions, such as postage costs and bus-stop queue problems, are discontinuous functions. $|x|$, which arises in 'distance away from' situations, is not smooth. Trigonometric, logarithmic, and exponential curves are not calculable by school arithmetic. Experiencing each of these situations takes students a new step away from elementary assumptions, a step towards an abstract understanding of functions, but we do not often see in textbooks coherent, explicit, comments about how these relate to the overarching idea of relations between variables. Instead they often appear fragmented, and subtle differences, such as being able to calculate values for quadratics but having to 'know' or 'find' or 'interpolate' values for sine curves, are often unremarked upon. In addition students get very strong prototypical images of what a graph of a function should look like (e.g. either linear or quadratic). In Even's research (1998), students claimed that these were the only possible graphs that could be drawn through some given points, even though they had met other functions.

There is little research on how students make sense of quadratics, yet the school curriculum expects a significant amount of time to be spent learning about them. Teppo and Esty (1994) analysed the kinds of questions asked in some textbooks about quadratics to identify typical progressions of understanding. Usually early questions vary the parameters $a$, $b$, and $c$ in the first of the notations offered above, and ask for rearrangement and factorisations. After these manipulations, a formula for calculating roots is introduced and practised. It seems obvious, with the availability of software, that students can then check what the graph looks like. Some teachers use zooming to find approximate roots before introducing formal methods, but use of the formula leads to questions about negative values for $b^2 - 4ac$. Some textbooks they looked at then moved on to introduce hidden quadratic structures, such as expressions in which the variable was $\sin\theta$ or $e^x$.

Using quadratics as 'the first non-linear function' introduces students to variable gradient, turning points, the relation between parameters and position and shape, the effects of scaling, and the meaning of roots of equations. In other words, quadratics provide the move from seeing functions as generating ordered pairs to functions being mathematical objects in their own right with certain properties (Hershkowitz and Bruckheimer, 1981). This could be given as one of the reasons for including them in the curriculum. When they are plotted from points these changes of perception are hard to make as students get bogged down in the calculations, but the use of software, focusing on construction and variation, makes the change more likely (Godwin and Beswetherick, 2003). Students who use scaling, scrolling, and zooming to explore tend to develop their own screen-related language for describing what they see. For example, students of Yerushalmy (1991) decided between linear and quadratic functions on the basis of the number quadrants they passed through. Conjectures like this can of course be challenged with non-examples, but teachers need first to recognise what students are thinking.

Regular movement between algebraic and graphical systems is a major teaching method. Each system presents particular problems and what is obvious in one is not necessarily obvious in the other (Leinhardt *et al.,* 1990). It is tempting to try to remove difficulties by treating functions as abstract, but there is much to be gained from using non-mathematical experience to inform the graphing process. Furthermore, one of the aims of mathematics education is to enable students to apply mathematical ideas in employment and other subjects, for which experience of modelling is a strong basis. Characteristics of the phenomenon being explored have to be understood with and through graphical and algebraic symbol systems.

Function notation, as all notations, presents its own problems. The use of $y =$ emphasises the dependent variable, but the $f(x)$ notation emphasises the independent variable(s) and is useful for multivariate situations. Sajka (2003) gives a range of student interpretations:

- $f$ as a label;
- $f(y)$ and $f(b)$ as different functions;
- $f(x)$ as the formula for a function;
- $f(3)$ means the function has value 3;
- $f(x + y) = f(x) + f(y)$;
- $f(y)$ is the ordinate;
- $f(x) = g(x)$ is an instruction to find an unknown;
- $f(x)$ is a graph;
- $f(x)$ means '$f$ multiplied by $x$'.

Similar confusions arise for particular functions, such as $\sin(x)$ being interpreted as 'sine multiplied by $x$', a confusion compounded by the use of $\sin^{-1}x$

on calculators as its inverse. We know of students for whom 'inverse' means the calculator shift button.

# Progression

## Being capable with functions

We shift now to the development of understanding through experience of function-related tasks. O'Callaghan (1998) identified five areas of expertise with functions: modelling, interpreting, translating between representations, treating them as objects, and acting on them with procedures. To achieve such expertise, students have to move between seeing functions as processes to seeing them as objects (Tall, 1992); from acting on them as objects to thinking about what is invariant under such actions; and from using prototypes to developing a wider repertoire of examples (Curcio, 1987; Schwarz and Hershkowitz, 1999).

They also need to handle formal definitions of function, but when and how formal definitions are introduced has to be thought about carefully. Elia *et al.* (2007) explored 16-year-old students' understanding of 'function' and found that knowing a formal definition did not help them identify functions when they were expressed as arrow diagrams or algebraically, but did help with graphs. They concluded that being given definitions at an early stage of understanding had not been useful. Many writers suggest delaying formal definitions (e.g. Confrey and Smith, 1994; Schwarz and Hershkowitz, 1999). Slavit (1997) focuses on properties rather than definitions, and compares descriptions of input/output processes to a growth-orientated view focusing on change. As we have seen, both of these are useful, but the ultimate aim is to achieve a view which permits: comparing and classifying functions; knowing their behaviour on the domain; and focusing on the effects of their parameters. Later they have to be combined to make compound functions.

Breidenbach *et al.* (1992) asked undergraduates for their perceptions of functions, and many of them described functions as actions, mainly ways to calculate a new set of values. For example:

- something that evaluates an expression in terms of $x$;
- an equation in which a variable is manipulated so that an answer is calculated;
- a combination of operations used to derive an answer.

Others gave a more abstract meaning in which the process is more like a 'black box' than an instruction to act:

- some sort of input being processed, a way to give some sort of output;
- an algorithm that maps an input into a designated output;
- an operation that accepts a given value and returns a corresponding value.

 www.nuffieldfoundation.org/functions-1

 www.nuffieldfoundation.org/functions-2

Students who held multiple meanings of 'function' were better able to see them as processes which could be combined as compound functions.

## Understanding covariation and rates of change

In calculus, students need to know how one variable changes in relation to another – this is a key property that lines and curves can represent. Understanding starts with qualitative judgements about slope, and also with linear growth patterns such as: when $x$ goes up by one, $y$ goes up by three, etc. Recursive adding-on patterns are then expressed as multiples of the independent variable. It is well-known that moving from term-to-term patterns in data tables to a global formula for the function is a difficult step involving good knowledge of notation, method of differences, or (as sometimes happens) a rote-learnt repertoire of methods of identifying functions from discrete differences. For understanding covariation, term-to-term analysis is a good basis for thinking about gradient and change over an interval and hence becoming more precise and quantitative about slope. Doorman and Gravemeijer (2009) found that using discrete graphs and focusing on changes between points helped students understand change over an interval. For example, in a growth situation students can be asked to describe the difference between two adjacent measurements. Success relates as much to understanding the context as to the methods used.

Full understanding of rate involves: an image of change in some quantity; coordination of images of two quantities, and an image of the simultaneous covariation (Thompson, 1994). In relation to this progression Carlson *et al.* (2002) identified levels of reasoning: coordination of variables; direction of change; expressing coordination of quantities; average rate over uniform increments;

and finally instantaneous rate. This, we suggest, provides a structure for appropriate task design and pedagogic prompts.

## Solving equations

Solving the kind of equations that arise in school can be seen as finding intersections of functions with each other or with $y = k$ (for some real $k$). Thus simultaneous equations can be seen as the intersection of two functions; solving quadratics as finding $x$-axis intercepts; solving linear equations as finding the $x$-coordinate of a particular point, and so on. If expressions in $x$ are usually graphed in classrooms, for example using hand-held technology, zooming methods are available and relevant for modelling realistic contexts. However, they are not generally accurate for mathematical purposes and treating 'solving' this way may avoid a powerful mathematical idea – that of equivalence of expressions – unless it is specifically emphasised (Kieran, 1984).

We first look at how students solve equations when only pencil and paper methods are available. These depend on understanding equivalence but can be reduced to procedures, either by teachers or textbooks or by efficient students who need to devise their own shorthand processes. For linear equations two main methods are traditional, both involving inverse operations (Adi, 1978):

- to see an expression as a sequence of operations which has to be undone;
- to see equality as a balance in which equilibrium has to be maintained, so if an operation on one side needs to be inverted, this also has to be applied to the other side, for example 'what you do to one side you do to the other'.

Adi claimed that the first method was easier for more students, particularly when using a 'cover-up' method. For example, in Figure 8.5, to solve: $6(x - 2) + 9 = 27$ first 'cover-up' $6(x - 2)$ and ask 'what number adds to 9 to give 27?' Then cover-up $(x - 2)$ and ask 'what number is multiplied by 6 to get 18?' and so on. The act of covering is isomorphic to inverting each operation in turn, as is done with the function machine approach, but where 'cover-up' looks for missing numbers, reversing a function machine uses knowledge of relations to identify inverses (Filloy and Rojano, 1989; Linchevski and Herscovics, 1996).

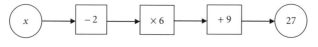

**Figure 8.5**

www.nuffieldfoundation.org/functions-1

 www.nuffieldfoundation.org/functions-3

Adi sees the balance metaphor as cognitively more challenging as it requires more imagination. Students who have been taught to 'do the same thing to each side' can develop, for themselves, the 'change the side, change the sign' shortened form (Lima and Tall, 2008). The same mnemonic can be derived from the function machine approach, but over-reliance can lead to misuse however it is derived.

One problem with balance is that the metaphor (if it makes sense at all in an age of digital weighing) does not extend to negative quantities so the images students have need to be extended (Vlassis, 2002). Boulton-Lewis *et al.* (1997a) tried providing concrete materials to model algebraic equations with 13- and 14-year-olds, to support the reasoning involved in solving linear equations using the balance model. Students devised their own reasoning using short-cuts which resembled the inverse reasoning method, but they found that the materials introduced extra stages and interpretation problems without helping the solution process, especially when negative numbers were involved. The balance metaphor has the advantage of working with both sides simultaneously, of maintaining the equation, rather than of thinking in a particular direction as is the case with changing sides, or covering-up, or function machine methods. Pirie and Martin (1997) claim that the difficulties in having an unknown on both sides and/or negative quantities are a result of pedagogy and not an inherent quality of equations. They describe how a teacher talked about quantities on each side of a fence having to be equal to avoid the physical problems of the balance image. He prepared students by previously solving missing number puzzles with unknowns on each side, so they needed a method that was more powerful than guesswork.

Another approach uses conservation of area in rearrangements of an area model for linear expressions. For example, if two areas of $3x$, each with one extra square, are known to be equal to one area of $7x$ with 3 squares removed, we can 'see' (in Figure 8.6) that two extra from the left-hand shapes are equal to $x - 3$ in the right-hand shape, so $2 = x - 3$ and hence $x = 5$.

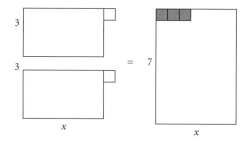

**Figure 8.6**

Students using this approach rarely make errors of transformation, whereas algebraic manipulations on their own are highly error prone (Perrenet and Wolters, 1994). Kieran (1981) and Steinberg *et al.* (1990) watched students find out if two expressions were equivalent, such as: $6x - 3$ and $3(2x - 1)$; $6x - 3$ and $6(x - 3)$; $6x - 3$ and $6(x - 1)$. They usually acted arithmetically, adding or subtracting numbers depending on the appearance of expressions. They had been taught to 'do the same to both sides' and some tried to apply this to the identification of equivalence. Very few tried to transform the expression on one side of the '=' to see if it could be made to match the other side, but more managed to identify equivalence by comparing terms, or by using simultaneous equation techniques of substitution. The lack of use of transformation suggests that what students learn to do in algebra lessons, such as collect like terms, factorise, expand, and so on, was not understood as generating alternative equivalent expressions. Equivalence can be demonstrated graphically of course, a particularly dramatic one being to graph $f(x) = \sin^2 x + \cos^2 x$ – and this could be used to further give meaning to algebraic manipulations.

Jones (2008) introduces the notion that one expression can be substituted for another with physical manipulation of on-screen objects. For example, if learners are given the fact that $24 = 6 \times 4$ they can then use '$6 \times 4$' instead of 24 wherever it occurs, and vice versa. Moving one expression into a position in another expression is the physical enactment of substitution. It is then a small step from doing this using numbers to doing this with algebraic expressions. Unfortunately we do not know the longer term effects of any of these different approaches on algebraic thinking and functional understanding, but Jones' focus on substitution, with its focus on 'can be exchanged for', seems more promising than the mnemonic 'change the side change the sign' in terms of extendable mathematical meaning, while the area approach, if maintained over time, seems to improve algebraic thinking (Fairchild, 2001).

## Expressing functions

Situations that generate linear functions ought to provide a context for relating expressions to graphs and situations. If students 'see' linear graphs arising from situations they already comprehend they may develop an understanding of relations between representations. However, the structure and behaviour of the situation has to be the first focus of this process. Wollman (1983) describes how natural language and the grammar of word problems can lead to errors in devising the algebraic expression. Students want to build algebraic expressions from left to right in response to words. But MacGregor and Stacey (1993) noticed something further: the mental model is more important than the wording of the problem. The main errors were due to failing to imagine the equality, and therefore constructing the reverse of the necessary relation, particularly in correspondence problems. For example, if a wooden pole needs 5 brackets to hold it up, then $b$ (the total number of brackets) is $5p$ (where $p$ is the number of poles), but many students will interpret the phrase '5 brackets' to mean $5b$ and write $p = 5b$.

In a teaching experiment over two lessons with 45 eight-year-olds, Warren and Cooper (2007) asked students to describe growth patterns. Tables of values appeared to draw students' attention away from the role of the position number as the variable. What helped included:

- making explicit connections between position number and result;
- explicit questions that focus on functional relation;
- presenting visual patterns not in sequence.

Even with these teaching methods some students found it hard, but the study did show that it is possible for many young children to develop functional thinking and express it verbally and symbolically in circumstances that draw their attention to the relevant connections (see also Blanton and Kaput, 2005).

Full functional thinking goes beyond inferring algebraic relations from data. Experts focus more on continuity, extreme values, subdomains on which they increase or decrease, symmetry, periodicity (Thomas *et al.*, 2010). Thus an expert might be considering which order of polynomials might best fit a dataset, where a novice might try to construct an expression for relating the coordinates of some data points.

# Teaching approaches

## Curriculum order

As with most major ideas in mathematics, there is no research that indicates better or worse curriculum ordering for these issues (Leinhardt *et al.*, 1990). Indeed, studies by Ainley and her colleagues with young children (1995, 2000) and Yerushalmy with functions of two variables (1997) confound any attempt to make claims about what can or should be taught at particular ages. Instead we think about what is made possible, and what is made more difficult, by different teaching approaches. For example, Hitt (1998, p. 12) identified components of progression towards full understanding:

- imprecise ideas about functions, probably mixing representations;
- identification of different representations;
- translation with preservation of meaning between representation systems;
- coherent joint use of representations in the solution of a problem.

This typography was relevant for students whose first experience of functions was as correspondences between sets of data, but it may not describe progress of students from representing relations between variables. Nevertheless Hitt's (1998) variety of pedagogic tasks (we present an adapted form) is a useful source of ideas and ends with an abstract understanding of functions:

- identify functions and their properties and definitions; identify domain and range;
- tabulation and graphing tasks; calculate values of the function at given points;
- translate between algebraic to graphical representations; physical context and pictorial modes;
- construct functions with given properties in algebraic and/or graphical form;
- decide the veracity of definitions of certain functions;
- identify functions which are equal or equivalent;
- operate with functions $(+, -, \times, o)$;
- use functions to prove or search for counter-examples.

Note that a modelling cycle is lurking in the middle of this list as construction and manipulation using different representations:

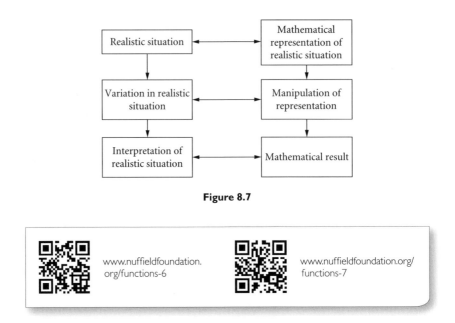

**Figure 8.7**

Students' developing understanding of connections between graphs, functions, realistic situations, and the special cases arising from solving equations, is only as strong as the curriculum organisation and teaching allows it to be. Leinhardt *et al.* (1990) identify two main kinds of action on functions expressed graphically: construction and interpretation. Each of these actions can be employed to predict, classify, and translate with functions. Most studies about learning functions, they say, are about interpretation rather than construction and yet through construction the conceptual understandings necessary for interpretation can be developed. There is not a rigid distinction between tasks for developing understanding and tasks for learning how to use functions, since the former informs the latter, and vice versa. We draw on their work by first looking at the interpretation.

### INTERPRETATION TASKS

We pointed out earlier that graphs can be interpreted as pictures, and Bell *et al.* (1987) developed a major teaching resource which tackled this (Swan, 1980). They provided graphs that had to be matched with activities and included possibilities that would lead to cognitive conflict. For example, they included a speed graph that looked like a fishing rod and line and a picture of an angler to tempt students to match them up – erroneously. Other obstacles to interpretation have already been indicated earlier: not understanding that every point on a line represents the underlying relationship; confusing height with slope; and misreading the scale. Leinhardt *et al.* found that most interpretation tasks in the literature required local quantitative attention, such as finding values or rates over an interval. Typical interpretation questions address:

- interpolation and extrapolation, especially the discrete/continuous confusion, expectations of continuity and shape, assumptions about joining points;
- particular events that show up on the graph: changes in quantity and rate, maxima and minima, zeroes, discontinuities;
- what noticeable features of the graph represent: changes of gradient, turning points, zeroes, intersections and intercepts;
- how *y* changes as *x* changes more generally.

## CONSTRUCTION TASKS

Plotting points was by far the most usual kind of construction task. We have said that this constrains understanding to (a) function as formula for calculating points and (b) the only relevant points being those plotted. Tasks that require construction of algebraic representations from properties of graphs are harder than other constructions as this requires an understanding of the whole relation rather than a point-wise view (Janvier, 1987).

Sfard and Linchevski (1994) analyse the demands on students of algebraic and graphical representations of functions and claim that the former is a more natural development, as it develops from arithmetic, where the latter requires new forms of understanding and interpretation. However, use of graphical representations in an integrated way, even with young children, contradicts this advice. There is considerable evidence that young children can express relations between variables through input-output models and tables of values when they understand the context, and when the independent variable progresses in unit steps (Martinez and Brizuela, 2006; Warren and Cooper, 2007).

Building graphs from situations is less common in most curricula than building from algebra, although this experience is necessary to become confident with modelling. Swan (1980) and Janvier (1981) share the view that students' experience of functions and their graphs should start with intuitive and qualitative models of well-known situations, rather than with abstract point-plotting. Older students might be asked to sketch functions by identifying their algebraic features, such as zeroes, asymptotes, turning points, and *y*-intercepts. The similarities and differences between drawing graphs from data points, from realistic situations, and from the properties of functions are rarely made explicit.

# Multiple representations

A consensus is developing around the extensive use of multiple representation software to enable students to connect algebraic and graphical systems, to learn about functions, and to learn to apply and recognise functions in realistic situations. Although in the past pencil and paper methods *had* to precede treating functions as objects and modelling with them, this is no longer the case and there are many long-term studies which demonstrate that even 8 and 9-year-olds who engage in modelling cycles using graphing software can, through these experiences, learn to graph without making the usual errors and also learn to express functions symbolically. Ruthven (1990) found that students between 11 and 13-years-old who used graphing calculators were more likely to connect symbolic terms, functional properties, and graphs than those who did not. The integrated and well-planned use of multiple representations to learn about functions, graphs, and equations makes a significant difference for students. Several medium and long term projects report similar findings and we shall summarise three of them.

### ACTIVE GRAPHING

Young children had personal access to laptops at all times from the age of 5 or 6 to use across the curriculum (Ainley, 1995; Ainley *et al.,* 2000). Over several years, researchers engaged students in 'active graphing' in which they collected data in experimental situations, gave children a suitable challenge, and ensured they had control of one variable (see also Chapter 6). Students entered data into spreadsheets and displayed it as graphs. Challenging questions required them to interpret the graph. Students experienced this approach several times up to the age of 11 and exhibited far fewer problems with interpreting graphs, and drawing their own graphs, than is generally the case. The research was based on the belief that the technicalities of producing graphs get in the way of meaningful use. Students aged 8 and 9 were even able to talk globally about graphs, even where they only consisted of data points, because the graphical image appeared instantaneously as one object. They developed a sense of what would be 'normal', or what could indicate a 'trend', and were able to correct data in order for the graph to make sense.

### VISUAL MATH

Over a three year course, Yerushalmy (1991, 2001) provided multiple representation experiences to enable 12 to 14-year-olds to develop a modelling perspective and to understand functions. She mapped progress over several months as students moved from being able to formulate problems mathematically, to mathematise quantities, to formulate suitable functions, and finally to form equations by comparing functions. This progress was scaffolded by sequences of tasks which were at first numerical, then used tables and graphs as parallel representations of variation and incremental change, and finally used global symbolic representations. Understanding developed throughout these modelling stages, starting with correspondence and variation, then moving through equations and inequalities and their manipulations, to exploring families of functions. In all the project materials (Yerushalmy *et al.*, 2002), the ideas of correspondence between variables and variation over intervals developed in parallel. Students needed numerical approaches for longer than Yerushalmy expected – they were reluctant to use graphing software as a tool until they felt confident with algebra, even though they had been shown linear and quadratic functions early to establish an initial repertoire. The emphasis was on formalising *after* the relations are understood, but notation itself still took a considerable time to learn.

## COMPUMATH

In a programme of work with 14 and 15-year-olds, Schwarz and Hershkowitz and their colleagues (1999) used multiple representation software to teach functions. Tasks were organised around problem situations which had meanings outside mathematics. The design team had, over 20 years, moved gradually from presenting formal definitions with abstract examples, through giving a definition plus exploratory tasks followed by formal procedures, and finally to using a chain of problems in which students constructed functions from informal descriptions – leaving formal definitions till later. In this latest version of the programme the main actions students can use are zoom, scale, and scroll. Students who have followed this version are good at recognising graphs of functions, even when they are only partial, can use prototype linear and quadratic functions to build new functions, and are also able to transform algebraic expressions appropriately. They focus on representations not as objects to be manipulated, but as having underlying meaning. Students were taught, using appropriate software:

- to see data points as belonging to a continuum;
- to compare and connect graphs;
- to compare alternative interpolations;

- to move points from tables to graphs and vice versa;
- to understand ordered pairs as instances of a function;
- to anticipate the shape of a graph from its algebraic expression;
- to use scaling to transform functions and identify change and invariance.

What else can a multiple representation approach achieve? Bloch (2003) used a similar approach with 17-year-old calculus students who did not have a strong sense of functions. She set tasks which required translations between multiple representations and allowed the properties of the underlying functions to be seen from several perspectives. Although the strongest students did no better with this approach than with more conventional approaches, average students improved their understanding of properties, finding formulae from graphs, exploring composite and inverse functions, changing graphs to change the algebra, and associated proofs. They developed a wider range of available functions to draw on than similar students taught without multiple representations.

There are several other smaller studies making similar claims. In some studies however it is not clear whether the aim is to use mathematics to find out more about realistic situations, or whether the situations are being used as vehicles to develop mathematical understanding. In the work of Bardini, Pierce, and Stacey (2004) functions were being used as a way to learn more about the power of algebra, so the role of letters as variables in functions enabled students to see them as variables in algebraic equations as well. Graphing calculators were used to solve linear equations, with the teacher showing how to write and transform algebraic notation when necessary for the task. The context, which was selling lemonade, provided the need for the construction of relations in conventional notations. The need for clarity of purpose is emphasised by Izsak and Findell (2005) who challenge the value of multiple representational approaches. They note that understanding how covariation is seen in a single representation takes time and deliberate attention to relating it to the original situation. But the proliferation of successful multiple representation approaches contradicts the need to learn one representation at a time.

Science provides many situations which could be used to provide a sequence which generates models which are mathematically similar (Michelson, 2006). This observation brings the issue of familiarity and abstraction to the fore. Students need several similar experiences for using the same functions in order to gain an abstract idea of the function and recognise its potential for future use. This is especially important for 'black box' functions which can then be compared and their properties understood in terms of the original phenomena.

You need a repertoire of functions in order to fit symbolic expressions to given graphs, and therefore you need such a repertoire to model realistic situations (Curcio, 1987).

## Functions of more variables

We said earlier that most studies do not show the longer term effects of different approaches, but we do know that in Yerushalmy's work (1997) students using graphing software were so competent at graphing and interpreting functions with one independent variable, and relations between quantities, that she developed tasks that involved more variables. This meant that two-dimensional graphing was no longer an option, and discussions of rate of change would have a much more complex meaning. She worked with 13-year-old students in their normal class and asked them to imagine an oil tanker being filled with several pipelines which were turned off one-by-one and then turned on again. They also had a car hire problem in which price per day and price per kilometre were combined for the total cost. With two variable data a point-by-point method had been used which needed information about rate of change over an interval. In the multivariate situations some of them developed the idea for themselves of controlling variables in order to understand the rates of change due to individual variables, and developed notation using three-dimensional drawings or towers of interlocking cubes. Multi-column tables enabled some to devise overall rules for the tanker contents or the total car hire cost, but in general students used arithmetical methods to solve particular cases rather than use the rules they developed. They had been enculturated into using 'f' to mean 'function' but apart from that had to decide how to write expressions involving two independent variables. They 'invented' multivariate expressions like:

$$f(n) = 100 + 5x \qquad f(n) + f(x) = 100n + 5x \qquad f(d,k) = 100n(d) + 5n(k)$$

Over several experiences with only one independent variable they had developed terminology that connected to what they could see graphically. They were so used to graphing that they would use such terminology even without drawing or having an available graph. They had become accustomed to using graphs as sense-making objects, and using the function notation to express the associated relations. They were also used to devising algebraic expressions and then testing them out graphically to see if they gave the expected values and behaviour. That enabled them to develop these sophisticated expressions at such a young age.

# Summary

Research has given many insights into the reasons why students find learning about functions hard, but no answers about 'best' methods or most effective curriculum ordering.

Of the range of ways available to learn how to solve algebraic equations, we have pointed out the limitations and confusions associated with each method, and the possible strengths. Methods which retain the notion of equality of expressions seem particularly powerful, and relate well to understanding the properties of functions.

There are many difficulties associated with graphs and graphing both realistic and algebraic data. Teachers need to consider that these may be problems of conceptual understanding rather than technical problems.

A full understanding of functions, both in pure mathematics and as tools for modelling, takes many years to develop and students need a variety of experiences. Students have to meet a wide range of functions: continuous and discrete; with and without time on the $x$-axis; smooth and not smooth. They have to meet functions that are not calculable as well as understanding those that are. They have to meet graphs that do not depict the underlying relations in obvious ways, as well as those that do. There are several distinct shifts of perspective to be made, and although there is no research about this, it is likely that the more teachers take these into account, the more probable it is that students will make them.

We do know, from research, that consistent use of multiple representation software over time can, with appropriate tasks and pedagogy, enable students to understand the concepts and properties associated with functions, and be able to use functions to model real and algebraic data.

# Where additional evidence is needed

This is an area of mathematics for which there is quite a lot of recent research, and much of it relates to the use of digital technology. Clearly, students need a coherent experience throughout school and over time in order to be brought to making the necessary shifts of perspective. For this reason, the knowledge practitioners can bring to the field is very valuable, such as:

- Construct maps of your students' developing understanding of quadratic functions.
- Find out where, in other subjects, students have to construct or interpret graphs,

and what kinds of graphs these are. What assumptions are made about their understanding of axes, scales, variables, accuracy, and other features? Can a coherent experience of graphing be developed in your school? Can science provide opportunities to generate non-linear identifiable functions?

- Devise questions to see if your students, after suitable teaching, can translate between graphs and algebra; tables of values and algebra; tables of values and graphs, and vice versa for all three. Try this with linear, quadratic, and unusual functions, such as step functions or |x|. Evaluate the findings and identify what theycan do most easily and what they find hard. What does this mean in terms of their past experience?

## Key readings

Leinhardt, G., Zaslavsky, O., and Stein, M. S. (1990). Functions, graphs and graphing: Tasks, learning, and teaching. *Review of Educational Research,* 1(1), 1–64.

This paper reviews research and describes the cognitive components of a good understanding of functions. It is especially good in providing information about the effects of different tasks.

Swan, M. (1980). *The language of functions and graphs.* Shell Centre for Mathematical Education. University of Nottingham.

This is a classic, well-designed, research-based collection of tasks which lead learners through the development of the notion of function by focusing on interpretation of graphs.

Yerushalmy, M. (1991). Students' perceptions of aspects of algebraic function using multiple representation software. *Journal of Computer Assisted Learning,* 7(1), 42–57.

This describes students' experiences and learning when they can switch between and control various representations, and when different representations are all easily available.

CHAPTER 9

# Moving to mathematics beyond age 16

## Introduction

This chapter illustrates that as students make the transition to mathematics beyond the age of 16 their mathematical experiences need to bring together the range of mathematical ideas encountered earlier on in their mathematical career; in other words, the ideas covered in the earlier chapters in this book. The new mathematical ideas encountered in the years beyond age 16 include, amongst other things, trigonometric functions, calculus and analysis, and statistical inference. These are amongst the topics that are at the heart of what is sometimes called 'higher' or 'senior' mathematics (leading to 'advanced' or 'formal' mathematics; see Edwards *et al.*, 2005; Tall, 1991, 2008). It is these topics that are addressed in brief in this chapter; a fuller treatment would need a whole new book.

As an example of what is involved in bringing together mathematical ideas, Watson (2009a, p. 5) uses the example of the topic of trigonometry to argue that:

Robust connections between and within earlier ideas can make it easier to engage with new ideas, but can also hinder if the earlier ideas are limited and inflexible. For example, learning trigonometry involves understanding: the definition of triangle; right-angles; recognizing them in different orientations; what angle means and how it is measured; typical units for measuring lines; what ratio means; similarity of triangles; how ratio is written as a fraction; how to manipulate a multiplicative relationship; what 'sin' (etc.) means as a symbolic representation of a function and so on. Thus knowing about ratios can support learning trigonometry, but if the understanding of 'ratio' is limited to mixing cake recipes it won't help much. To be successful, students have to have had enough experience to be fluent, and enough knowledge to use methods wisely.

That students have to be in the right position to embark on 'higher' or 'senior' mathematics is echoed in the joint position statement on the teaching of calculus from the Mathematical Association of America (MAA) and the National Council of Teachers of Mathematics (NCTM) (2012). This statement posits that the goal of school mathematics education 'should not be to get students into and through a course in calculus by twelfth grade [when students are 18] but to have established the mathematical foundation that will enable students to pursue whatever course of study interests them when they get to college'. Given that the precursors of calculus are ideas of algebra, ratio, similarity, measure, decimals, graphs, slope/rate/gradient, functions, right-angled triangles, and circle geometry (the latter through the use of radians as a measure), what the MAA and NCTM point to is the danger that students can get short-changed in their preparation in algebra, geometry, trigonometry, and other mathematical topics, in order to stay on a fast track to calculus. It is the avoiding of such short-changing that is at the heart of all the earlier chapters of this book.

The third focus that we consider in this final chapter is that of statistical inference; something that builds on earlier stochastic ideas such as standard deviation, covariation, and probability distribution. Across these three areas of mathematics (trigonometric functions, calculus/analysis, statistical inference) there are ideas of mathematical modelling at this level that bring together ideas of algebra and functions, proportionality, measure (especially compound measure), and sophisticated forms of reasoning about data.

Beyond the scope of this chapter are the myriad uses of mathematics in the workplace, such as in business and finance, in engineering, and in design. Much of this mathematics is touched upon in the mathematics curriculum at this level but it seems there is often not the curriculum flexibility to be able to do much more. For example, in finance there are mathematical ideas of algebra, proportionality, functions, data, probability, measures, discrete and decision mathematics, hypothesis testing, and the analysis of risk; the mathematics in engineering and physics encompasses mechanics and kinematics which in turn involve at the very least: algebra, proportionality, measure, spatial reasoning (including vectors), probability, and functions. Meanwhile in design there are mathematical ideas of proportionality, measure, spatial reasoning, vectors, statistics, functions, dimensionality, similarity, and symmetry.

For reasons of practicality in a book of this size, this chapter has to concentrate on selected central ideas; in this case on trigonometric functions, on calculus and analysis, and on statistical inference. In doing so, we review the nature of each area of mathematics and draw on existing relevant research that addresses how these aspects of mathematics can be taught and learnt.

## Trigonometric functions

### The nature of trigonometry in school

Trigonometry (from the Greek for triangle measure) began as a branch of geometry, both for the practical task of surveying land and for astronomical calculations. Trigonometric functions are pervasive in many parts of pure and applied mathematics. For example, trigonometric expressions occur in the solutions of algebraic equations such as $x^3 = 1$, and power series involve the inverse tangent function.

This means that in the later years of school, students learn about trigonometric functions, trigonometric equations, trigonometric identities, and the solving of problems involving oblique triangles (ones that are not right-angled) using what are commonly known as the Laws of Sines and Cosines, as well as extending right-angled triangle calculations to 3D situations. Beyond this there is the use of complex numbers (numbers of the form $x + iy$, where $x$ and $y$ are real numbers and $i = \sqrt{-1}$ ) in trigonometric expressions and in Euler's formula $e^{i\phi} = \cos\phi + i\sin\phi$, where $e$ is the base of natural logarithms.

### Groundwork prior to trigonometric functions

In preparation for learning about trigonometric functions, students need to be familiar with a range of ideas across algebra and geometry. Analyses of the mathematical ideas involved in trigonometric situations, even fairly simple ones, show that students who cannot appreciate a ratio given as a number, or manipulate a multiplicative relation, are at a clear disadvantage to those who can (Weber, 2005, 2008). This could underlie why it can be tempting to provide shortcuts to remembering how to use various transformations of the formulae for sine, cosine, and tangent, and to remembering which ratio involves which sides. One example of this is the popular mnemonic 'SOH-CAH-TOA' for the trigonometric ratios.[1] A further assumption about students in terms of pre-trigonometry is that they can readily identify right-angled triangles in atypical situations; this is not always the case (Byers, 2008).

On top of all this, trigonometry is often the first context in which students meet functions that are not polynomials in $x$ and that are represented using a name. For students who have used the $f(x)$ notation this involves replacing $f$ by 'sin' but if their previous experience is solely of notations like equations '$y =$' or like mappings '$x \rightarrow$', then using a name for a function can be mystifying.

Trigonometry may also be the first context in which the notion of inverse function has to be discussed. The shift of understanding from a procedural approach to the resolution of problems involving triangles (using degrees) to working with functions (using radians) is difficult for learners (Martinez-Sierra, 2008; Orhun, 2001). The available evidence is that this is not solely about learning a different notation (Byers, 2008; Pritchard and Simpson, 1999). The lesson is that students who have been asked to relate a sine curve to a cosine curve may not get far if their concept image of such functions is dominated by SOH-CAH-TOA.

## Challenges in learning about trigonometric functions

Existing research reported at conferences (for example, Byers, 2008; Delice and Monaghan, 2005; Kendal and Stacey, 1996), and reports of teaching ideas in professional journals (for instance, Kemp, 2009; Steer *et al.*, 2009a, 2009b, 2009c; Weber, 2005), confirm that the initial stages of learning about trigonometric functions can be challenging for students. This is partly because of the need to relate diagrams of triangles to numerical relationships and then manipulate the symbols involved; another reason is because students have trouble viewing trigonometric relationships as functions particularly if most of their previous experience has been using formulae to solve right-angled triangles (Blackett and Tall, 1991; Breidenbach *et al.*, 1992; Weber, 2005, 2008).

[1] This stands for: 'sine, opposite, hypotenuse, cosine...etc.'

## Teaching approaches

An experienced head of department said to us of trigonometry:

The boundaries of the topic seem to be clearly defined, at least at first glance. It elicits such comments from my colleagues as 'I'm saving it until the start of next term'. It has pupil status; this is the stuff their older siblings and parents remember. It involves using the, as yet, mysterious buttons on the calculator and finally being in the know. My department likes teaching trigonometry. Each experienced teacher has a tried and tested progression through the theory and routines.

While the illusion of clear boundaries can lead to narrow procedural teaching, a variety of approaches to introducing trigonometric ideas are taken in practice (Watson, 2009b).

- *Similarity*: students find that the ratio of pairs of sides in sets of similar triangles is the same.
- *Functions*: students plot values of the height of a point moving on a unit circle and relate this to the right-angled triangle it forms with the radius and x-axis.
- *Multiplier*: students relate heights of various right-angled triangles to a fixed-length hypotenuse using 'sine' on a calculator.
- *Ratio*: students use ratios to solve scaling problems and are then introduced to the names for these ratios.
- *Procedural resolution of triangles*: students are presented with trigonometric ratios as tools to use, alongside Pythagoras, to find missing values.
- *Exploration of calculator functions*: students explore what the function does by controlling input variables and plotting data or recording output in a spreadsheet.

Each of these approaches goes some way towards addressing the conceptual problems described above; some of the approaches emphasise the multiplicative relationship, while others emphasise the spatial meaning. All except the procedural approach aim to give experience of what 'sine' is, and what it does before introducing its name and notation.

Kendal and Stacey (1996) query the longer term effects of different methods. They used 12 questions to test students who had been taught using a unit circle method and those taught a procedural method. Most questions required the calculation of a nominated side length in a right angle triangle, involving: choosing the correct function; formulating the equation; transforming and solving the equation. Some questions required solving equations of the same structure that arises in trigonometry, that is equations of the type $a = bc$ or $b = \frac{a}{c}$.

Their conclusion was that a procedural method was more efficient in enabling students to solve these kinds of question, but the questions were indeed limited to those that required procedural knowledge. It is not clear, from their study, which students could adapt their knowledge to the more mathematically-sophisticated roles of trigonometry. If the unit circle method merely provides a diagram from which the right-angled triangle has to be extracted and dealt with separately, then all it has achieved is extra transformations to carry out. Yet if the unit circle is used to present 'sine' as the multiplier which, when applied to a radius, evaluates height, no transformation is necessary. Lakoff and Núñez (2000) argue that the unit circle provides a 'natural' understanding of trigonometric ideas which can be developed beyond 360 degrees, a further advantage of using this image.

Other writers have reported success in engendering robust understandings of trigonometric ideas, beyond what is needed for solving right-angled triangles, by combining unit circle, similarity, and exploratory approaches. For example, Steer, de Vila and Eaton (2009a, 2009b, 2009c) report a series of lessons designed by Jeremy Burke which move from a mathematical description of trigonometry, through a set of classroom activities (using dynamic geometry tools to permit exploration) which combine ratio and unit circle methods. Students developed their own accounts of trigonometry, using these tools, over four lessons. Further teaching experiments of this type would be useful.

## The value of learning trigonometric functions

Learning trigonometry takes place through an accumulation and overlapping of experiences over several years – and possibly several teachers – and depends on students' trajectories of experience rather than on one-off classroom events.

Neither professional publications nor research reports about learning trigonometry point to problems with students' prior understandings of angles, ratio, or algebraic manipulation – all ideas that are difficult to learn (see earlier chapters). Perhaps it is only those students who have overcome typical problems with these earlier concepts who progress to be taught trigonometric functions.

We prefer to take the view that such issues with prior understandings exist, but that working on trigonometry provides a context for developing a relational (as compared to instrumental) understanding of the component concepts. For example, understandings of ratio can develop through work on similar right-angled triangles and the notion of sine, cosine, and tangent as multipliers;

understanding of angle as a measure of turn develops through extension beyond 90 degrees.

Understanding algebraic manipulation in trigonometry is probably limited because of the notation problems associated with the functions, and this is why it may be tempting to provide mnemonics and transformation aids, or memorised formulae, rather than using the opportunity to work on the algebra of functions in a meaningful context. Given that a major value of trigonometry is in the use of functions in higher mathematics, sciences, and engineering, teaching needs to take such limitations into account.

## Where evidence is needed

As noted above, existing research specifically on the teaching and learning of trigonometric functions is relatively limited. Hence there remain a wide range of issues on which evidence is needed including:

- the affordances and limitations of different ways of introducing trigonometric ideas and developing these towards an understanding of trigonometric functions;
- how trigonometric understanding develops over time and multiple experiences;
- how the teaching and learning of trigonometric functions can be enhanced by the use of digital technologies such as data-loggers to provide data in order to model phenomena such as periodic motion (for example, a pendulum).

# Calculus and analysis

## The nature of calculus and analysis

In general terms, a calculus refers to any system of calculation guided by the symbolic manipulation of expressions. One example is propositional calculus, a part of mathematical logic. In mathematics at the level under consideration in this chapter, the calculus being considered is the study of change and is commonly divided into two major branches: differential calculus (or differentiation) and integral calculus (or integration). What links the two is the remarkable fact that differentiation and integration are inverses, as encapsulated by the Fundamental Theorem of Calculus. Historically, the calculus was developed independently by Newton and by Leibniz in the seventeenth century and centred on 'infinitesimals'. It was formulated as a way of finding the slopes of curves, the areas under curves, and of calculating minima and maxima. Nowadays, at school level, differential calculus is considered as being concerned with things like velocity and acceleration which might involve calculating the slope of a curve and the rate of change of that slope, and with optimisation which might involve calculating minima or maxima. Integral calculus encompasses computations involving area, volume, arc length, and centre of mass, amongst other things.

Mathematical analysis, usually referred to simply as 'analysis', is the branch of pure mathematics that encompasses differential and integral calculus but extends to cover measure, limits, infinite series, and analytic functions. Analysis is the mathematics of functions, with real analysis and complex analysis being the two broad subdivisions that deal with real-valued and complex-valued functions, respectively.

## The calculus in the 16–19 age range

As Tall (1993) explains, for students beyond the age of 16 the calculus can mean different things in different countries. In some countries, what might be termed informal calculus is taught to relevant students in the 16–19 age range. This is concerned with informal ideas of rate of change and the rules of differentiation, and with integration as the inverse process used for calculating areas and volumes. This is followed at university level with formal real analysis which includes the rigorous *epsilon–delta* ($\varepsilon$–$\delta$) definitions of limits, continuity, differentiation, Riemann integration, together with formal deductions of theorems such as mean-value theorem, and the fundamental theorem of calculus. Other countries leave the calculus to university level and, at that point, might begin formally with the *epsilon–delta* definition, or may begin with some aspects of informal calculus before moving on to formal analysis. Still other countries (Greece being an example) begin formal analysis in the latter stages of school, prior to university. The approaches used, including the level of mathematical rigour, the representations employed (geometric, numeric, symbolic), and the individual topics covered, can vary greatly from country to country.

Whatever approach is taken, and whenever it occurs that students begin to study the calculus, their success depends on their previous experience and current knowledge of both algebra and geometry. Of particular import is students' knowledge and ideas about functions and their capability of manipulating algebraic expressions; equally important are their ideas of ratio, similarity, measure, slope/rate/gradient, right-angled triangles, and circle geometry as it relates to radians. All this entails students not only being able, for instance, to interpret the graphs of functions such as simple polynomials, but also knowing about rational functions, trigonometric functions, the relationship between powers and logarithms, and so on. Such knowledge is vital to being able to formulate the numerical approximation of the slope of a function, or to construct an expression for differentiating or integrating 'from first principles' (also known as the '*delta* method'), or to interpret results. When students experience difficulties with the fundamental theorem of calculus, Thompson (1994, p. 229) argues, these stem from 'impoverished concepts of rate of change and from poorly developed and poorly coordinated images of functional covariation and multiplicatively-constructed quantities'.

## The challenge of teaching and learning calculus and analysis

Whatever way the calculus is approached in school or at university, the need to improve calculus courses and its teaching has been for some time a theme for research from around the world. A major reason for the research focus on calculus education is the widespread recognition that in teaching and learning the calculus, there is not only a wide range of ideas from algebra and geometry that need to be brought together, but the new ideas that are generated through this bringing together are cognitively complex as well (Robert and Speer, 2001; White and Mitchelmore, 1996). These new ideas include the mathematics of limits and infinity, both of which provide major challenges for students (Davis and Vinner, 1986; Williams, 1991). Other challenges relate to the area and volume computations, to understanding the convergence of sequences, to relating the slope of the tangent to the limit of the rate of increment of the function (that is, seeing the tangent line as the limit of secant lines, visually and analytically), and so on.

Attempts to avoid this complexity through students being offered solely some procedural rules mostly lead to misapplication and inability of students to apply methods appropriately (e.g. Orton, 1983a, 1983b). An example of misapplication can occur when students who have been taught a rule for polynomial differentiation end up trying to differentiate $e^x$ as $xe^{x-1}$. Other examples of misapplication include students diving into differentiation to find attained extrema on a closed interval without considering the endpoints, or integrating to find area without considering the shape of the graph.

At this point, we need to distinguish between the learning focus of students who, without some background understanding, may resort to rule-based learning even when something more meaningful is offered, and the teaching focus of teachers who act as if their role is to simplify the subject-matter and hence only offer procedures. Although it can happen that students end up muttering 'bring down the $n$ and reduce the power by one' to differentiate $x^n$ with respect to $x$, that does not mean it should be taught that way.

Recognition of the difficulties that arise for students who are taught the calculus beginning with the *epsilon–delta* definition of limits has led to other teaching approaches being developed. These non-definitional methods include 'real-world calculus in which intuitions can be built enactively using visuo-spatial representations, through the numeric, symbolic and graphic representations in elementary calculus and on to the formal definition–theorem–proof–illustration approach of analysis' (Tall, 1996, p. 294). One important factor in such developments in teaching has been the increasing availability of computing technology, especially the provision of visuo-spatial representations and the use of inter-related numeric, symbolic, and graphical representations, so that changes to one representation (such as the symbolic equation) can be replicated instantly in a parallel (say, graphical) representation (e.g. Heid, 1988; Tall *et al.*, 2008). We say more about this in the next section.

## Teaching approaches

An alternative to beginning the calculus with the *epsilon–delta* definition of limits is to introduce differentiation through finding the gradient of the tangent to the curve at a point by finding the numerical limiting value of the gradient of the secant drawn through the point and an adjacent point as the second points converges on the first point (sometimes called 'from first principles' or the 'delta method'). This avoids problems with *epsilon–delta* but introduces problems of its own. Cavanagh (1996), for instance, reveals that the approach is characterised by complicated algebraic manipulations and the central idea of the limiting process can be lost on students.

One way to change the teaching approach has been to use what can be called the 'locally-straight approach' to the calculus (Tall, 1985). In this approach, differential calculus is based around the idea of magnifying (or zooming in on) a part of a graph of a function to see it approximate to a straight line whose slope can be measured, as in Figure 9.1. The idea then is to move along the graph to see the changing slope.

Such a 'locally straight' approach has been adopted in a variety of ways and in various places. In general, the teaching approach is for students to study the graph of suitable functions and use appropriate computer software to sketch the graphs of the changing slope of the respective curves. In this way,

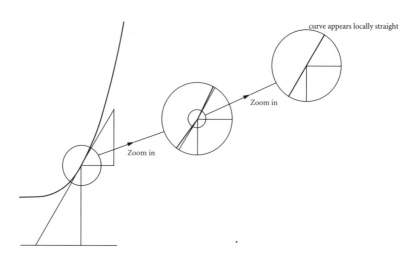

curve appears locally straight

Zoom in

Zoom in

**Figure 9.1** Magnifying (or zooming in on) a graph to see it approximate to a straight line.

the students might observe that the derivative of $x^2$ is $2x$ of $x^3$ is $3x^2$, and they might conjecture the general result. This can later be extended to 'seeing' that the derivative of sin$x$ is cos$x$ (by observing the relative shapes of the graphs), and perhaps to finding the numerical value of $k$ for which the derivative of $k^x$ is $k^x$ (by choosing values of $k$ between 2 and 3). Through the 'locally-straight' approach, the value of the slope is seen dynamically as a function in its own right. Importantly, the focus on 'local straightness' should be combined from the outset with a consideration of 'non-local straightness'; in other words with the idea of non-differentiability. Overall, this 'locally-straight' approach to the calculus appears to provide a suitable initial stage to the calculus that can later lead not only to practical calculus applications but also to formal *epsilon–delta* analysis, or to non-standard analysis using infinitesimals (Tall, 1996).

While 'local straightness' involves imagining a graph being highly magnified to 'see' how steep it is at a chosen point, another teaching approach that is used in some places uses the idea of 'local linearity'. This latter approach focuses on finding a linear function that is the 'best linear approximation' to the graph at the chosen point. In contrast to using 'local straightness', symbolism is used from the outset. In this way 'local linearity' places more demands on students' prior understandings than does the notion of 'local straightness' (for more on this, see Tall, Smith, and Piez, 2008).

The idea of 'local straightness' is one example of the use of graph-plotters (be these computer-based or a capability of hand-held calculators). Such technologies can be used to show how changing a parameter in a function changes the shape of the graph and also its gradient function. This allows new questions to be posed that may be beyond the reach of students using conventional paper-and-pencil methods, and which can necessitate deeper understanding. An example of such a task might be: 'design a cup by finding some parabola which can enclose a capacity of 175 ml with height between 10 and 12 cm'.

While the idea of 'local straightness' is being used in the teaching of differential calculus, the idea of 'accumulation' (of a quantity described by its rate of change) is being suggested for use in the teaching of integral calculus. As Thompson and Silverman (2008, p. 49) explain, the idea of accumulation 'both grows out of and contributes to a coherent understanding of rate of change. When something changes, something accumulates. When something accumulates, it accumulates at some rate. To understand rate of change well therefore means that one sees accumulation and its rate of change as two sides of a coin. Thus, students' success in the integral calculus can begin in middle school if rate of change is taught substantively'. Yerushalmy and Swidan (2012) illustrate the use of a dynamic and multi-representational computer environment in the early stages of learning of ideas of accumulation by two high-school calculus students (aged 17) who had not yet learned any other integral-related ideas, definitions, or procedures.

Technology may also provide a vehicle for hands-on explorations that might lead towards more formal definition–theorem–proof approaches. For example, Quesada *et al.* (2008) describe a short-term teaching experiment during which college students were taught how to approach *epsilon–delta* problems through the use of a graphing calculator. They found that 'students who received the graphical instruction for two days performed significantly better on all but one of the conceptual items on the post-test, as compared to the students who received traditional instruction and with those who received the graphical instruction for only one day' (p. 95). More recently, Swinyard (2011) provided details (from a 10-week-long small-scale teaching experiment) of the experience of two students neither of whom had previously seen the conventional *epsilon-delta* definition of limit. Through a series of tasks with access to graphing calculators they were both able to 'characterize limit in a manner synonymous to that of the conventional $\varepsilon$–$\delta$ definition' (p. 111). For more details of research on using technology in calculus teaching, see Ferrara *et al.* (2006) and Tall *et al.* (2008).

These teaching approaches (including 'local straightness' and 'accumulation') exploit the potential of computer environments to handle multiple representations (graphical, numeric, symbolic) of mathematical functions. This raises the question of whether students significantly under the age of 16, given appropriate technologies, might be able to engage with the mathematics of change and variation that is the focus of the calculus (Stroup, 2003). This is something that Kaput and his colleagues (e.g. Kaput and Schorr, 2008; Roschelle *et al.*, 2000) began investigating. This work is continuing with new collaborators (e.g. Roschelle *et al.*, 2010). In this initiative, the technology provides dynamic graphs and animated cartoon worlds to engage elementary, middle, and high school students in reasoning about the relationships between position, velocity, and acceleration. The aim is to construct a strand of the school mathematics curriculum that is neither a simplified symbolic calculus course, nor a typical exploration of linear functions and the related notions of rate and ratio. Rather, the idea is for a completely new curriculum (see Chapter 8 for more about functions).

## Where evidence is needed

Existing research on the teaching and learning of the mathematics of change and variation is pointing to calculus-related ideas being something to be studied across many years of school, perhaps even from the earliest grades. This might entail a strand of the school mathematics curriculum that is neither a simplified version of symbolic calculus, nor an exploration of linear functions and related notions of rate and ratio. Evidence is needed on what such a strand of school mathematics might look like but we think that a strong emphasis on relations between quantities might provide the baseline, as it does for algebra and for proportional reasoning.

Research on the use of digital technologies in the teaching and learning of the calculus is continuing to point to the affordances of such technologies to provide insights for students. Evidence is needed about ways in which digital technologies can be used to link measurable events that are experienced as meaningful by students to ways of representing those events in formal mathematical terms. Again, this points to a baseline need for familiarity with manipulating relations between quantities.

## Formal statistical inference

### The nature of formal statistical inference

The emphasis in Chapter 6 on informal inferential reasoning was motivated by two concerns: first, evidence is suggesting that an exploratory data analysis approach might support interpretation of data in ways that are of profound importance to the layperson. This motivation is based on the contribution of informal inferential reasoning to statistical literacy. The second motivation is a conjecture that experience in informal inference might facilitate better understanding of classical formal statistical inference, which, as shown below, presents significant conceptual difficulties. Formal statistical inference is widely used in social science and in other disciplines as a major method for testing conjectures about patterns in data.

In most scientific approaches to making sense of phenomena, a key, possibly counter-intuitive, driver in statistical thinking is that it is acausal (Ramsey, 1999), depending rather on a mastery of variability than on cause and effect (Gould, 2004). In a deterministic model of some phenomena, variation from the model can typically be conceptualised as error, such as when measuring independent and dependent variables as part of an experiment to see how the one causes change in the other. The error is perceived as an unfortunate accident to be avoided as far as possible. In contrast, in a stochastic model, variability provides the essential data in the form of a distribution (Wild, 2006). The shape of that distribution might provide key insights into the patterns in the data or it might provide a sense of the reliability of perceived effects or associations.

In fact, statisticians have many ways of thinking about variability (Borovcnik, 2005). Data are perceived as comprising two components (each of which could be further subdivided): component 1 might be described *inter alia* as the signal, the main effect or the explained variation; component 2 is then described respectively as the noise, the error or deviation, or the residual (see Chapters 6 and 7). The random element of the model that is associated with the variation is aligned with the second component.

Formal statistical inference is made up of a (very large) set of tools and methods for deciding whether the second component is sufficiently small to justify a hypothesis that the first component is in some sense real, and not just a result of the vagaries of chance. Two such methods are hypothesis testing and the computation of confidence intervals. Both sets of techniques depend fundamentally on the notion that variation is inevitable and, in uncertain situations, error may arise through measurement, guessing and estimating, variation in machine performance, and so on. In this sense, error is not a mistake but an inevitable consequence of uncertainty and is a resource as far as statistical analysis in concerned. Such a perspective on error may conflict with how errors are often perceived (as mistakes) in lower school. We note, however, that certain types of teaching, such as in the active graphing approach discussed in earlier chapters, may help to undermine this negative view of errors. As well as this strong grasp of randomness, hypothesis testing and the computation of confidence intervals require a good appreciation of modelling, distribution, sampling and the Law of Large Numbers, all seen as key concepts in Chapter 6.

## Formal statistical inference in the 16–19 age range

As mathematics students move through ages 16–19, they are likely to encounter formal inference. If they continue to study to degree level, they are likely to meet formal statistical inference whether in mathematics, a science, or a social science.

Early encounters with formal statistical inference typically begin with reinforcing representations of location (with most emphasis on the mean), dispersion (standard deviation) and covariation (scattergraphs). The focus on representations is likely to continue through the introduction of various types of probability distribution such as binomial, uniform, Poisson, and most importantly the normal distribution. Finally, more formal methods for measuring correlation are likely to be introduced prior to the introduction of hypothesis testing and confidence intervals.

Most students will have encountered a similar diet of representations when learning about data handling lower down the school and may well continue to struggle with the underlying purpose of being introduced to many different representations and techniques. Chapter 6 proffered an alternative approach centred on informal exploration of data through the data-handling cycle and perhaps such an approach might have lessened the range of difficulties reported in students' understanding of formal inference.

## The challenge of teaching and learning formal statistical inference

Certainly it seems that many of the problems associated with heuristic thinking (Kaplan and Du, 2009), the notion of random variable (Miller, 1998), the Law of Large Numbers (Finch, 1998), standard deviation (delMas and Liu, 2005), distribution (Noll *et al.*, 2010), sampling and variability (Hjalmarson *et al.*, 2011; Noll, 2011), and sample versus population mean (Aquilonius, 2005; Zendrera, 2010) persist into higher education. In fact, there is remarkably little research directly on the 16–19 age range so in effect the best we can do for teachers of this age is interpolate between findings for younger and older students.

A review of the above papers seems to suggest that the prevailing teaching methods fail to facilitate the construction of powerful meanings for the most powerful statistical ideas with the result that at college level many students are introduced to formal statistical inference without a robust appreciation of the key underpinning ideas.

Covariation and correlation are given far more emphasis from age 16. Previously, work would typically be limited to the use of scattergraphs, often to present data. Misconceptions of correlation have been classified (Batanero *et al.*, 1996; Estepa and Sánchez Cobo, 2001, 2003) as causal/deterministic (believing that correlations indicate causal or functional relationships), local (utilising only particular parts of the data provided in a contingency table), or unidirectional (taking negative correlation as indicating independence). Sotos *et al.* (2009) tested 279 undergraduates, mostly first year, across various disciplines and added the non-transitivity misconception to the categories of misunderstandings about correlation. This may need some explanation. Most relations encountered in lower school are transitive. Thus, if $a = b$ and $b = c$ then $a = c$. Or, if $a$ divides by $b$, and $b$ divides by $c$, then $a$ divides by $c$. In everyday life, many relations are not transitive. So, if Dave is Annie's parent and Annie is Tony's parent, Dave is not Tony's parent. Correlation is an example of a non-transitive relation in mathematics. So, the occurrence of lung cancer may be correlated with smoking, and smoking may be correlated with bad breath, but the occurrence of lung cancer may not be correlated with bad breath. Nevertheless, about half of the students in this study responded to the test questions as if Pearson's correlation coefficient were transitive.

Hypothesis testing is a central element of formal statistical inference. Empirical studies have identified misconceptions of hypothesis testing. A basic misunderstanding, found for example by Vallecillos (1995), was that many students will regard a low $p$-value as absolute evidence of the alternative hypothesis. It is also common for students to confuse practical significance and statistical significance, not appreciating therefore that a statistically significant result may have little consequence or importance in practical terms. For example, with a large sample, it may be possible to show a small correlation is statistically significant but the existence of a small correlation may be of no importance and can only be judged from contextual knowledge (Gliner *et al.*, 2002). Vallecillos and Batanero (1996) point to the lack of clarity about null and alternative hypothesis, with students often choosing the incorrect hypothesis. Later in the hypothesis testing process, students often confuse the significance level and the $p$-value (Falk, 1986).

Another aspect of formal inference that has attracted the attention of researchers is that of confidence intervals, seen by some as a suitable response to the widely misunderstood hypothesis test. Fidler (2006) found that use of confidence intervals rather than hypotheses tests reduced incidence of the confusion between statistical and practical significance but was nevertheless associated with many additional misunderstandings. Furthermore, several studies of researchers' understanding of confidence intervals (for example, Cumming *et al.*, 2004) suggests that there are conceptual hurdles with confidence intervals as well and it is not clear that a switch to the use of confidence intervals will in itself improve the general level of understanding of statistical inference. Canal and Gutiérrez (2010) investigated students and university statistics experts and found that both groups sometimes suggested that confidence intervals include sample means or other single statistics rather than possible values of a parameter, such as the population mean.

## Teaching approaches

This catalogue of failure to understand formal statistical inference merely highlights the problem and the possible failure of prevailing pedagogic methods. It is possible to speculate that teaching approaches at lower ages that emphasise exploratory data analysis and informal inference might result in more robust understanding of formal inference but at present there is no research to confirm that conjecture. There are however a few studies that do provide some indications of what might be more effective as a teaching approach for older students.

Corredor (2008) explored the teaching and learning of ANOVA through two comparative methods. One intervention used data analysis tools and focused on authentic situations. The other used simulations and focused on formal aspects of probability. The use of simulations proved more effective. Garfield *et al.* (2010) reported on an introductory statistics course which was designed around model-eliciting activities (MEAs). MEAs offer experience of authentic statistical problems based on real data. Two examples are: (i) determining whether an iPod shuffle tool generates random playlists; (ii) designing a SPAM filter for email messages. Although the project has not yet been formally assessed and evaluated, early qualitative results suggest that the students are responding to these tasks by engaging in deep statistical thinking.

It would appear at face value that the Garfield *et al.* study contradicts that of Corredor but it is important to recognise the very different aims. Corredor's aim was to improve understanding of ANOVA. When the teacher's aim is specific to a particular learning outcome, it is likely that contextual information will merely hinder understanding of that concept or skill so that simulating the situation so that students access the logic behind the statistical idea or procedure may well be effective. In contrast, Garfield *et al.* appear to be seeking a broader appreciation of the nature of statistical analysis and in that sense their approach mirrors that proposed for younger students in Chapter 6. With that broader aim, students need to engage with the full data handling cycle at whatever age they might be and conceptual information is essential. Of course, it would be entirely possible to identify points in a course based on statistical exploration where a specific focus, perhaps based on simulation, would help students to grasp a more formal understanding of key statistical concepts.

## Where evidence is needed

There is almost no research in formal statistical inference in 16–19 year olds. As informal inference becomes more widespread as a teaching approach for younger students, opportunities will emerge to research the impact of such a pedagogy on the 16–19 age range and the ongoing transition to university level.

The research methods used to explore formal statistical inference focus almost exclusively on cataloguing misconceptions. Arguably, this is of limited use to teachers, who need also to know what students do understand in order to plan for progress that builds on that knowledge. One exception is the study by Garfield *et al.* (2010) where they explored a particular teaching approach based around MEAs in higher education. There is a need for far more evidence about which teaching approaches support deeper understanding of formal statistical inference.

# Teaching for conceptual growth through powerful ideas

In the cases of trigonometry and calculus we have shown that understanding depends on drawing together multiple strands of knowledge and understanding that have been developed across and throughout school mathematics. Within a strong whole-school approach to those components we can imagine students experiencing a coherent and progressive conceptual development of ideas. However, it is also the case that deeper understanding of earlier ideas can come about by seeing how they arise in more complex contexts, which explains how it is that many learners are successful even in situations where they have a less than coherent curriculum. In the case of statistical inference a new kind of reasoning has to be learnt; and this new reasoning can sometimes runs counter to their everyday reasoning, and also to other mathematical reasoning based on exclusive categories. It can also run counter to their beliefs about situations, so there is a strong need for teachers to know how to support students in new forms of thought.

It is vital, therefore, that students build knowledge on well-prepared ground and that teachers recognise how that ground needs to be prepared. We are not talking here about emotional and social commitment to mathematics, although these are important, but about the preparation of the mind. We therefore return in this final section to the mathematical themes we mentioned in Chapter 1.

## Teaching approaches

Again and again we found in the research that students who do no more than adopt rules for action can misapply them and over-generalise their use, so we see a need for students to understand why particular actions are relevant for particular mathematical situations by also experiencing what does not work, and why. Software, particularly that which offers several mathematical representations (e.g. graphical, numeric, symbolic), can enable rapid exploration of different situations so that students are not dependent on teachers and textbooks for examples and non-examples but can generate such things for themselves.

## Conceptual growth

Confusion, contradiction, and misapplication have to be regarded as inevitable states of the learning mind as new classifications are being understood and new symbolisations manipulated. With the most complex ideas, it is multiple experiences from several points of view, and over time, that are required to achieve a workable understanding. This is especially true of multiplicative reasoning, which arises in many of our chapters. However, this is only one of several forms of reasoning that are necessary to learn mathematics to higher levels: deductive (geometry), structural (algebra), statistical, probabilistic, estimating, predicting, hypothesising, axiomatic, and transformational reasoning. Some of these depend on knowledge, some on ways of seeing problems, and some on logical reasoning methods. As we have seen in this chapter in the case of statistical inferential reasoning, learners can find it hard to recognise and adopt new ways of thinking particularly if such ideas are left solely as implicit.

## Powerful ideas

Across the chapters in this book we located variability, proportionality, similarity, symmetry, linearity, measure, dimensionality, representations, prediction, accuracy, discrete/continuous number, and transformation as key powerful ideas in mathematics. When we work with groups of teachers it takes time for these to come to the surface; it is more usual for teachers to mention pattern-spotting, problem-solving, learning methods, and some other task-types at first. This may be because the curricula or textbooks list the content for them and they have to think about how to focus students' minds on particular content. Again and again in our reading of the research, the same central ideas emerge as key to later understanding and most of them are either about how quantities relate to each other, or how such relations can be formalised and represented so they can be transformed, adapted and hence used.

# REFERENCES

Abrahamson, D. and Wilensky, U. (2007). Learning axes and bridging tools. *International Journal of Computers for Mathematics Learning*, 12(1), 23–55.

Accascina, G. and Rogora, E. (2006). Using Cabri3D diagrams for teaching geometry. *International Journal for Technology in Mathematics Education*, 13(1), 11–22.

ACME (Advisory Committee on Mathematics Education) (2011). *Mathematical needs: The mathematical needs of learners*. London: ACME.

Adi, H. (1978). Intellectual development and reversibility of thought in equation solving. *Journal for Research in Mathematics Education*, 9, 204–213.

Ainley, J. (1991). Is there any mathematics in measurement? In D. Pimm and E. Love (Eds), *Teaching and learning school mathematics* (pp. 69–76). London: Hodder and Stoughton.

Ainley, J. (1995). Re-viewing graphing: Traditional and intuitive approaches. *For the Learning of Mathematics*, 15(2), 10–16.

Ainley, J. (2000). Exploring the transparency of graphs and graphing. In T. Nakahara and M. Koyama (Eds), *Proceedings of the 24th Conference of the International Group for the Psychology of Mathematics Education* (Vol. 2, pp. 9–16). Hiroshima, Japan.

Ainley, J., Nardi, E., and Pratt, D. (2000). Towards the construction of meanings for trend in Active Graphing. *The International Journal of Computers for Mathematical Learning*, 5(2), 85–114.

Ainley, J., Pratt, D., and Hansen, A. (2006). Connecting engagement and focus in pedagogic task design. *British Educational Research Journal*, 32(1), 23–38.

Alsina, C. and Nelsen, R. (2006). *Math made visual: Creating images for understanding mathematics*. Washington, DC: The Mathematical Association of America.

Anderson, J. R. and Fincham, J. M. (1994). Acquisition of procedural skills from examples. *Journal of Experimental Psychology: Learning, Memory, and Cognition*, 20, 1322–1340.

Aquilonius, B. C. (2005). How do college students reason about hypothesis testing in introductory statistics courses? Unpublished PhD Thesis, University of California, Santa Barbara, USA.

Arcavi, A. (1994). Symbol sense: The informal sense-making in formal mathematics. *For the Learning of Mathematics*, 14(3), 24–35.

Armstrong, B. and Larson, C. (1995). Students' use of part-whole and direct comparison strategies for comparing partitioned rectangles. *Journal for Research in Mathematics Education*, 26(1), 2–19.

ASE-Nuffield (2010). *The language of measurement: Terminology used in school science investigations*. Hatfield: Association for Science Education and the Nuffield Foundation.

Askew, M. and Wiliam, D. (1995). *Recent research in mathematics education 5–16*. London: HMSO.

Atiyah, M. (2001). Mathematics in the 20th century. *American Mathematical Monthly*, 108(7), 654–666.

Austin, L. B. (1995). Comment on 'Progression in measuring'. *Research Papers in Education*, 10(2), 175–176.

Baccaglini-Frank, A. and Mariotti, M. A. (2010). Generating conjectures in dynamic geometry: The maintaining dragging model. *International Journal of Computers for Mathematical Learning*, 15(3), 225–253.

Bakker, A. and Derry, J. (2011). Lessons from inferentialism for statistics education. *Mathematical Thinking and Learning*, 13(1 & 2), 5–26.

Bakker, A. and Gravemeijer, K. P. E. (2004). Learning to reason about distribution. In J. Garfield and D. Ben-Zvi (Eds), *The challenge of developing statistical literacy, reasoning and thinking*, pp. 147–168. Dordrecht, The Netherlands: Kluwer.

Bakker, A., Wijers, M., Jonker, V., and Akkerman, S. F. (2011). The use, nature and purposes of measurement in intermediate-level occupations. *ZDM: The International Journal on Mathematics*, 43(5), 737–746.

Banchoff, T. (1990). *Beyond the third dimension: Geometry, computer graphics, and higher dimensions*. New York: Scientific American Library.

Barbeau, E. J. (1988). Which method is best? *Mathematics Teacher*, 81, 87–90.

Bardini, C., Pierce, R., and Stacey, K. (2004). Teaching linear functions in context with graphics calculators: Students' responses and the impact of the approach on their use of algebraic symbols. *International Journal of Science and Mathematics Education*, 2(3), 353–376.

Barrett, J. and Clements, D. (2003). Quantifying path length: Fourth-grade children's developing abstractions for linear measurement. *Cognition and Instruction*, 21(4), 475–520.

Barrett, J., Clements, D., Sarama, J., *et al.* (2012). Evaluating and improving a learning trajectory for linear measurement in elementary grades 2 and 3: A longitudinal study. *Mathematical Thinking and Learning*, 14(1), 28–54.

Batanero, C., Estepa, A., Godino, J. D., and Green, D. R. (1996). Intuitive strategies and preconceptions about association in contingency tables. *Journal for Research in Mathematics Education*, 27(2), 151–169.

Batanero, C., Navarro-Pelayo, V., and Godina, J. D. (1997). Effect of implicit combinatorial model on combinatorial reasoning in secondary school pupils. *Educational Studies in Mathematics*, 32(2), 181–199.

Batenaro, C., Biehler, R., Maxara, C., *et al.* (2005). Using simulation to bridge teachers' content and pedagogical knowledge in probability. *Paper presented at ICMI Study 15, The Professional Education and Development of Teachers of Mathematics.* Aguas de Lindoia, Brazil.

Battista, M. T. (1999). Fifth graders' enumeration of cubes in 3D arrays: Conceptual progress in an inquiry-based classroom. *Journal for Research in Mathematics Education*, 30(4), 417–48.

Battista, M. T. (2003). Understanding students' thinking about area and volume measurement. In D. Clements and G. Bright (Eds), *Learning and teaching measurement* (pp. 122–142). Reston, VA: NCTM.

Battista, M. T. (2004). Applying cognition-based assessment to elementary school students' development of understanding of area and volume measurement. *Mathematical Thinking and Learning*, 6(2), 185–204.

Battista, M. T. (2006). Understanding the development of students' thinking about length. *Teaching Children Mathematics*, 13(3), 140–147.

Battista, M. T. (2007). The development of geometric and spatial thinking. In F. Lester (Ed.), *Second handbook of research on mathematics teaching and learning* (pp. 873–885), Charlotte, NC: NCTM/Information Age Publishing.

Battista, M. T. (2010). Thoughts on elementary students' reasoning about 3D arrays of cubes and polyhedra. In Z. Usiskin, K. Andersen, and N. Zotto (Eds), *Future curricular trends in school algebra and geometry* (pp. 183–202). Charlotte, NC: Information Age Publishing.

Battista, M. T. and Clements, D. H. (1996). Students' understanding of three-dimensional rectangular arrays of cubes. *Journal for Research in Mathematics Education*, 27(3), 258–292.

Baturo, A. and Nason, R. (1996). Student teachers' subject matter knowledge within the domain of area measurement. *Educational Studies in Mathematics*, 31(3), 235–268.

Bednarz, N. and Janvier, B. (1996). Emergence and development of algebra as a problem solving tool: Continuities and discontinuities with arithmetic. In N. Bednarz, C. Kieran and L. Lee (Eds), *Approaches to algebra: perspectives for research on teaching* (pp. 115–136). Dordrecht: Kluwer.

Bednarz, N., Kieran C., and Lee, L. (1996). Approaches to algebra: Perspectives for research on teaching. In N. Bednarz, C. Kieran and L. Lee (Eds), *Approaches to algebra: perspectives for research on teaching* (pp. 3–14). Dordrecht: Kluwer.

Behr, M., Harel, G., Post, T., and Lesh, R. (1992). Rational number, ratio, proportion. In D. A. Grouws (Ed.), *Handbook of research on mathematics teaching and learning* (pp. 296–333). New York: Macmillan.

Bell, A. (1996). Algebraic thought and the role of manipulable symbolic language. In N. Bednarz, C. Kieran and L. Lee (Eds), *Approaches to algebra: Perspectives for research on teaching* (pp. 151–154). Dordrecht: Kluwer.

Bell, A., Brekke, G., and Swan, M. (1987). Diagnostic teaching: 5, graphical interpretation teaching styles and their effect. *Mathematics Teaching*, 120, 50–57.

Bell, S. (1999). *A beginner's guide to uncertainty in measurement.* Teddington: NPL.

Ben-Chaim, D., Lappan, G., and Houang, R. T. (1985). Visualizing rectangular solids made of small cubes: Analyzing and effecting students' performance. *Educational Studies in Mathematics*, 16(4), 389–409.

Ben-Chaim, D., Fey, J. T., Fitzgerald, W. M., *et al.* (1998). Proportional reasoning among 7th grade students with different curricular experiences. *Educational Studies in Mathematics*, 36(3), 247–273.

Ben-Zvi, D. (2004). Reasoning about variability in comparing distributions. *Statistics Education Research Journal*, 3(2), 42–63.

Ben-Zvi, D. (2006). Scaffolding students' informal inference and argumentation. In A. Rossman and B. Chance (Eds), *Proceedings of the Seventh International Conference on Teaching of Statistics* (CD-ROM). Voorburg, The Netherlands: International Statistical Institute.

Biggs, J. B. and Collis, K. F. (1982). *Evaluating the quality of learning: The SOLO taxonomy.* New York: Academic.

Bishop, J. W. (1997). Middle school students' understanding of mathematical patterns and their symbolic representations. *Paper presented at the annual meeting of the American Educational Research Association*, Chicago, Illinois.

Blackett, N. and Tall, D. O. (1991). Gender and the versatile learning of trigonometry using computer software. In F. Furinghetti (Ed.), *Proceedings of the 15th conference of the International Group for the Psychology of Mathematics Education* (Vol. 1, pp. 144–151). Assisi, Italy.

Blanton, M. and Kaput, J. (2005). Characterizing a classroom practice that promotes algebraic reasoning. *Journal for Research in Mathematics Education*, 36(5), 412–446.

Bloch, I. (2003). Teaching functions in a graphic milieu: What forms of knowledge enable students to conjecture and prove. *Educational Studies in Mathematics*, 52(1), 3–28.

Boero, P. (2001). Transformation and anticipation as key processes in algebraic problem solving. In R. Sutherland, T. Rojano, A. Bell, and R. Lins (Eds), *Perspectives on school algebra* (pp. 99–119). Dordrecht: Kluwer.

Booth, L. R. (1981). Child-methods in secondary mathematics. *Educational Studies in Mathematics*, 12(1), 29–41.

Booth, L. R. (1984). *Algebra: Children's strategies and errors*. Windsor, Berks: NFER-Nelson.

Borovcnik, M. (2005). Probabilistic and statistical thinking. In M. Bosch (Ed.), *Proceedings of the Fourth Congress of the European Society for Research in Mathematics Education* (pp. 485–507). Sant Feliu de Guixols, Spain.

Boulton-Lewis, G., Cooper, T., Atweh, B., Pillay, H., Wilss, L., and Mutch, S. (1997a). Processing load and the use of concrete representations and strategies for solving linear equations. *Journal of Mathematical Behavior*, 16(4), 379–397.

Boulton-Lewis, G., Wilss, L., and Mutch, S. (1997b). Analysis of primary school children's abilities and strategies for reading and recording time from analogue and digital clocks. *Mathematics Education Research Journal*, 9(2), 136–151.

Brandom, R. (2002). Overcoming a dualism of concepts and causes: A unifying thread in empiricism and the philosophy of mind. In R. Gale (Ed.), *The Blackwell Guide to Metaphysics* (pp. 263–281). Hoboken, NJ: Blackwell.

Brantlinger, A. (2011). Rethinking critical mathematics: A comparative analysis of critical, reform, and traditional geometry instructional texts. *Educational Studies in Mathematics*, 78(3), 395–411.

Breidenbach, D., Dubinsky, E., Hawks, J., and Nichols, D. (1992). Development of the process conception of function. *Educational Studies in Mathematics*, 23(3), 247–285.

Brown, M. (1981). Place value and decimals. In K. Hart (Ed.), *Children's understanding of mathematics 11–16* (pp. 48–65). London: Murray.

Brown, M., Blondel, E., Simon, S., and Black, P. (1995a). Progression in measuring. *Research Papers in Education*, 10(2), 143–170.

Brown, M., Blondel, E., Simon, S., and Black, P. (1995b). 'Progression in measuring': Response by the authors to the comments by Owen van den Berg and Lydia Austin. *Research Papers in Education*, 10(2), 177–179.

Brown, M., Jones, K., and Taylor, R. (2003). *Developing geometrical reasoning in the secondary school: Outcomes of trialling teaching activities in classrooms* (a report to the QCA). London: QCA.

Bryant, P. (2009). Paper 5: Understanding space and its representation in mathematics. In T. Nunes, P. Bryant, and A. Watson, *Key Understandings in Mathematics Learning*. London: Nuffield Foundation. Retrieved 17.05.12 from: http://www.nuffieldfoundation.org/key-understandings-mathematics-learning [accessed 4 August 2012].

Burger, W. and Shaughnessy, J. (1986). Characterizing the van Hiele levels of development in geometry. *Journal for Research in Mathematics Education*, 17(1), 31–48.

Burny, E., Valcke, M., and Desoete, A. (2009). Towards an agenda for studying learning and instruction focusing on time-related competencies in children. *Educational Studies*, 35(5), 481–492.

Byers, P. (2008). Examining trigonometric representations as a source of student difficulties. *Proceedings of the 11th Conference on Research in Undergraduate Mathematics Education* (RUME 2008). San Diego, California.

Canal, G. Y. and Gutiérrez, R. B. (2010). The confidence intervals: A difficult matter, even for experts. *Proceedings of the Eighth International Conference on the Teaching of Statistics.* Retrieved on 10.2.2012 from: http://www.stat.auckland.ac.nz/~iase/publications/icots8/ICOTS8_C143_CANAL.pdf [accessed 4 August 2012].

Cannizzaro, L. and Menghini, M. (2006). From geometrical figures to definitional rigour: Teachers' analysis of teaching units mediated through van Hiele's theory. *Canadian Journal of Science, Mathematics and Technology Education,* 6(4), 369–386.

Carlson, M., Jacobs, S., Coe, E., *et al.* (2002). Applying covariational reasoning while modeling dynamic events: A framework and a study. *Journal for Research in Mathematics Education,* 33(5), 352–378.

Carpenter, T., Corbitt, M., Kepner, H., *et al.* (1981). Decimals: Results and implications from national assessment. *Arithmetic Teacher,* 28(8), 34–37.

Carraher, D. (1993). Lines of thought: A ratio and operator model of rational number. *Educational Studies in Mathematics,* 25(4), 281–305.

Castro, C. S. (1998). Teaching probability for conceptual change. *Educational Studies in Mathematics,* 35(3), 233–254.

Cavanagh, M. (1996). Student understandings in differential calculus. In P. Clarkson (Ed.), *Technology in mathematics education: Proceedings of the 19th annual conference of the Mathematics Education Research Group of Australasia* (pp. 107–114). Melbourne: MERGA.

Cavanagh, M. (2007). Year 7 students' understanding of area measurement. In K. Milton, H. Reeves, and T. Spencer (Eds), *Mathematics: Essential for learning, essential for life* (Proceedings of the 21st Biennial Conference of the Australian Association of Mathematics Teachers, pp. 136–143). Adelaide: AAMT.

Cave, P. (2007). *Primary school in Japan: Self, individuality and learning in elementary education.* Abingdon, UK: Routledge.

Chang, K.-L., Males, L. M., Mosier, A., and Gonulates, F. (2011). Exploring US textbooks' treatment of the estimation of linear measurements. *ZDM: The International Journal on Mathematics,* 43(5), 697–708.

Chazan, D. (1993). High school geometry students' justifications for their views of empirical evidence and mathematical proof. *Educational Studies in Mathematics,* 24(4), 359–387.

Chick, H. L. and Watson, J. M. (2001). Data representation and interpretation by primary school students working in groups. *Mathematics Education Research Journal,* 13(2), 91–111.

Christou, C., Jones, K., Mousoulides, N. and Pittalis, M. (2006). Developing the 3DMath dynamic geometry software: Theoretical perspectives on design. *International Journal for Technology in Mathematics Education,* 13(4), 168–174.

Clark, M., Berenson, S., and Cavey, L. (2003). A comparison of ratios and fractions and their roles as tools in proportional reasoning. *Journal of Mathematical Behavior,* 22(3), 297–317.

Clement, J. (1985). Misconceptions in graphing. *Proceedings of the 9th Conference of the International Group for the Psychology of Mathematics Education* (Vol. 1, pp. 369–375). Utrecht, Netherlands.

Clements, D. H. (2003). Teaching and learning geometry. In J. Kilpatrick, W. G. Martin, and D. Schifter (Eds), *A research companion to principles and standards for school mathematics* (pp. 151–178). Reston, VA: National Council of Teachers of Mathematics.

Clements, D. H. and Battista, M. T. (1990). The effects of Logo on children's conceptualizations of angle and polygons, *Journal for Research in Mathematics Education*, 21(5), 356–371.

Clements, D. H. and Battista, M. T. (1992), Geometry and spatial reasoning. In D.A. Grouws (Ed.), *Handbook of research on mathematics teaching and learning*. New York: Macmillan.

Clements, D. H. and Bright, G. (Eds) (2003). *Learning and teaching measurement*. Reston, VA: NCTM.

Clements, D. H., Battista, M. T., and Sarama, J. (2001). *Logo and geometry* (Journal for Research in Mathematics Education, Monograph Number 10). Reston, VA: NCTM.

Clements, D. H., Battista, M. T., Sarama, J., and Swaminathan, S. (1996). Development of turn and turn measurement concepts in a computer-based instructional unit. *Educational Studies in Mathematics*, 30(4), 313–337.

Cobb, P. A. (1999). Individual and collective mathematics development: The case of statistical data analysis. *Mathematical Thinking and Learning*, 1(1), 5–44.

Confrey, J. (1995). Student voice in examining "splitting" as an approach to ratio, proportions and fractions. In L. Meira and D. Carraher (Eds), *Proceedings of the Nineteenth International Conference for the Psychology of Mathematics Education* (Vol. 1, pp. 3–29). Recife, Brazil.

Confrey, J. and Smith, E. (1994). Exponential functions, rates of change and the multiplicative unit. *Educational Studies in Mathematics*, 26(2–3), 135–164.

Confrey, J. and Smith, E. (1995). Splitting, covariation and their role in the development of exponential functions. *Journal for Research in Mathematics Education*, 26(1), 66–86.

Cooper, C. and Dunne, M. (2000). *Assessing children's mathematical knowledge*. Buckingham, UK: Open University Press.

Cooper, H. M. (1998). *Synthesizing research: A guide for literature reviews*. Thousand Oaks, CA: Sage Publications.

Cooper, M. (1992). Three-dimensional symmetry. *Educational Studies in Mathematics*, 23(2), 179–202.

Cooper, M. and Sweller, J. (1989). Secondary school students' representation of solids. *Journal for Research in Mathematics Education*, 20(2), 202–212.

Corredor, J. A. (2008). Learning statistical inference through computer-supported simulation and data analysis. Unpublished PhD Thesis, University of Pittsburgh.

Cramer, K. and Post, T. (1993). Connecting research to teaching proportional reasoning. *Mathematics Teacher*, 86(5), 404–407.

Cumming, G., Williams, J., and Fidler, F. (2004). Replication, and researchers' understanding of confidence intervals and standard error bars. *Understanding Statistics*, 3(4), 199–311.

Curcio, F. (1987). Comprehension of mathematical relationships expressed in graphs. *Journal for Research in Mathematics Education*, 18(5), 382–393.

Curcio, F. R. (1989). *Developing graph comprehension*. Reston, VA: National Council of Teachers of Mathematics.

Curry, M. and Outhred, L. (2005). Conceptual understanding of spatial measurement. In P. Clarkson, A. Downton, D. Gronn, M. Horne, A. McDonough, R. Pierce, and A. Roche (Eds), *Building connections: Theory, research and practice* (Proceedings of the 27th annual conference of the Mathematics Education Research Group of Australasia, Melbourne, pp. 265–272). Sydney: MERGA.

Davis, J. (2007). Real-world contexts, multiple representations, student-invented terminology, and y-intercept. *Mathematical Thinking and Learning*, 9(4), 387–418.

Davis, P. J. (1993). Visual theorems. *Educational Studies in Mathematics*, 24(4), 333–144.

Davis, R. B. (1985). ICME 5 Report: Algebraic thinking in the early grades. *Journal of Mathematical Behavior*, 4(2), 195–208.

Davis, R. B. and Vinner, S. (1986). The notion of limit: Some seemingly unavoidable misconception stages. *Journal of Mathematical Behavior*, 5(3), 281–303.

Davydov, V. (1990). *Types of generalisation in instruction: Logical and psychological problems in the structuring of school curricula*. Reston, VA: NCTM.

De Bock, D., Verschaffel, L., and Janssens, D. (1998). The predominance of the linear model in secondary school students' solutions of word problems involving length and area of similar plane figures. *Educational Studies in Mathematics*, 35(1), 65–83.

De Bock, D., Verschaffel, L., and Janssens, D. (2002). The effects of different problem presentations and formulations on the illusion of linearity in secondary school students. *Mathematical Thinking and Learning*, 4(1), 65–89.

de Villiers, M. (1991). Pupils' needs for conviction and explanation within the context of geometry. *Pythagoras*, 26(1), 18–27.

de Villiers, M. (1998). To teach definitions in geometry or teach to define? In A. Olivier and K. Newstead (Eds), *Proceedings of the 22nd Conference of the Psychology of Mathematics Education* (Vol. 2, pp. 248–255), Stellenbosch, South Africa.

Del Grande, J. (1990). Spatial sense. *Arithmetic Teacher*, 37(6), 14–20.

Delice, A. and Monaghan, J. (2005). Tool use in trigonometry in two countries. Paper presented to *Fourth Conference of the European Society for Research in Mathematics Education* (CERME4). Sant Feliu de Guixols, Spain.

delMas, R. and Liu, Y. (2005). Exploring students' conceptions of the standard deviation. *Statistics Education Research Journal*, 4(1), 55–82.

Dickson, L. (1989). The area of a rectangle. In K. Hart, D. Johnson, M. Brown, *et al.* (Eds), *Children's mathematical frameworks 8–13: A study of classroom teaching.* London: NFER-Nelson.

Dierdorp, A., Bakker, A., Eijkelhof, H., and van Maanen, J. (2011). Authentic practices as contexts for learning to draw inferences beyond correlated data. *Mathematical Thinking and Learning*, 13(1 & 2), 132–151.

Doorman, L. and Gravemeijer, K. (2009). Emergent modeling: Discrete graphs to support the understanding of change and velocity. *ZDM: The International Journal on Mathematics Education*, 41, 199–211.

Dossey, J. A. (1997). Defining and measuring quantitative literacy. In L. A. Steen (Ed.), *Why numbers count: Quantitative literacy for tomorrow's America* (pp. 173–186). New York: College Entrance Examination Board.

Dougherty, B. J. (2008). Measure up: A quantitative view of early algebra. In J. J. Kaput, D. W. Carraher, and M. L. Blanton (Eds), *Algebra in the early grades* (pp. 389–412). Mahweh, NJ: Erlbaum.

Dreyfus, T. and Eisenberg, T. (1990). Symmetry in mathematics learning. *ZDM: The International Journal on Mathematics Education*, 22(2), 53–59.

Duval, R. (1995). Geometrical pictures: kinds of representation and specific processings. In R. Sutherland and J. Mason (Eds), *Exploiting mental imagery with computers in mathematics education* (pp. 142–157). Berlin: Springer.

Duval, R. (1998). Geometry from a cognitive point of view. In C. Mammana and V. Villani (Eds), *Perspectives on the teaching of geometry for the 21st Century: an ICMI study* (pp. 37–52). Dordrecht: Kluwer.

Edwards, B. S. and Ward, M. B. (2004). Surprises from mathematics education research: Student (mis)use of mathematical definitions. *American Mathematical Monthly*, 111(5), 411–424.

Edwards, B. S., Dubinsky, E., and McDonald, M. A. (2005). Advanced mathematical thinking. *Mathematical Thinking and Learning*, 7(1), 15–25.

Elia, I., Panaoura, A., Eracleous, A., and Gagatsis, A. (2007). Relations between secondary pupils' conceptions about functions and problem solving in different representations. *International Journal of Science and Mathematics Education*, 5(3), 533–556.

Engel, J. and Sedlmeier, P. (2005). On middle-school students' comprehension of randomness and chance variability in data. *ZDM: The International Journal on Mathematics Education*, 37(3), 168–177.

English, L. and Sharry, P. (1996). Analogical reasoning and the development of algebraic abstraction. *Educational Studies in Mathematics*, 30(2), 135–157.

Erfjord, I. (2011). Teachers' initial orchestration of students' dynamic geometry software use: consequences for students' opportunities to learn mathematics. *Technology, Knowledge and Learning*, 16(1), 35–54.

Estepa, A. and Sánchez Cobo, F. T. (2001). Empirical research on the understanding of association and implications for the training of researchers. In C. Batanero (Ed.),

*Training researchers in the use of statistics* (pp. 37–51). Granada, Spain: International Association for Statistical Education and International Statistical Institute.

Estepa, A. and Sánchez Cobo, F. T. (2003). Evaluación de la comprensión de la correlación y regresión a partir de la resolución de problemas [Evaluation of the understanding of correlation and regression through problem solving]. *Statistics Education Research Journal*, 2(1), 54–68.

Even, R. (1998). Factors involved in linking representations of functions. *Journal of Mathematical Behavior*, 17(1), 105–121.

Fairchild, J. (2001) Transition from arithmetic to algebra using two-dimensional representations: A school based research study. *Papers on Classroom Research in Mathematics Education*, Centre for Mathematics Education Research, University of Oxford.

Falk, R. (1986). Misconceptions of statistical significance. *Journal of Structural Learning*, 9, 83–96.

Fast, G. R. (1999). Analogies and reconstruction of probability knowledge. *School Science and Mathematics*, 99, 230–240.

Ferrara, F., Pratt, D., and Robutti, O. (2006). The role of technologies for the teaching of algebra and calculus. In A. Gutiérrez and P. Boero (Eds), *Handbook of research on the psychology of mathematics education: Past, present and future* (pp. 237–273). Rotterdam, The Netherlands: Sense Publishers.

Fey, J. (1990). Quantity. In L. A. Steen (Ed.), *On the shoulders of giants: New approaches to numeracy* (pp. 61–94). Washington, U.S.A.: National Academy Press.

Fidler, F. (2006). Should psychology abandon p-values and teach CIs instead? Evidence based reforms in statistics education. In *Proceedings of ICOTS-7, Seventh International Conference on Teaching Statistics*. International Association for Statistical Education. Salvador, Brazil.

Filloy, E. and Rojano, T. (1989). Solving equations: The transition from arithmetic to algebra. *For the Learning of Mathematics*, 9(2), 19–25.

Finch, S. (1998). Explaining the Law of Large Numbers. *Proceedings of the Fifth Conference on the Teaching of Statistics*. Retrieved on 10.2.2012 from: http://www.stat.auckland.ac.nz/~iase/publications/2/Topic6n.pdf [accessed 4 August 2012].

Fischbein, E. (1975). *The intuitive sources of probabilistic thinking in children*. Holland: Reidel Publishing Company.

Fischbein, E. (1993). The theory of figural concepts. *Educational Studies in Mathematics*, 24(2), 139–162.

Fischbein, E. and Nachlieli, T. (1998). Concepts and figures in geometrical reasoning. *International Journal of Science Education*, 20(10), 1193–1211.

Fischbein, E., Deri, M., Nello, M., and Marino, M. (1985). The role of implicit models in solving verbal problems in multiplication and division. *Journal for Research in Mathematics Education*, 16(1), 3–17.

Fischbein, E., Nello, M. S., and Marino, S. M. (1991). Factors affecting probabilistic judgements in children in adolescence. *Educational Studies in Mathematics*, 22(6), 523–549.

Font, V., Godino, J., Palans, N., and Acevedo, J. (2010). The object metaphor and synecdoche in mathematics classroom discourse. *For the Learning of Mathematics*, 29(1), 15–19.

Forno, D. M. (1929). Notes on the origin and use of decimals. *Mathematics Newsletter*, 3(8), 5–8.

Freudenthal, H. (1971). Geometry between the devil and the deep sea. *Educational Studies in Mathematics*, 3(3 & 4), 413–435.

Freudenthal, H. (1973). *Mathematics as an educational task*. Dordrecht, The Netherlands: Kluwer.

Freudenthal, H. (1983). *Didactical phenomenology of mathematical structures*. Dordrecht, The Netherlands: Reidel.

Friedlander, A. and Lappan, G. (1987). Similarity: Investigations at the middle grades level. In M. Lindguist and A. Shulte (Eds), *Learning and teaching geometry, K-12* (pp. 136–143). Reston, VA: NCTM.

Friedman, W. J. and Laycock, F. (1989). Children's analog and digital clock knowledge. *Child Development*, 60(2), 357–371.

Fuys, D., Geddes, D., and Tischler, R. (1988). *The van Hiele model of thinking in geometry among adolescents* (JRME Monograph Number 3). Reston, VA: NCTM.

Gal, H. and Linchevski, L. (2010). To see or not to see: Analyzing difficulties in geometry from the perspective of visual perception. *Educational Studies in Mathematics*, 74(2), 163–183.

Galuzzi, M., Neubrand, M., and Laborde, C. (1998). The evolution of curricula as indicated by different kinds of change in geometry textbooks. In C. Mammana and V. Villani (Eds), *Perspectives on the teaching of geometry for the 21st Century: an ICMI study* (pp. 204–222). Dordrecht, The Netherlands: Kluwer.

Garfield, J. and Ben-Zvi, D. (Eds) (2005). *The challenge of developing statistical literacy, reasoning and thinking*. Dordrecht, The Netherlands: Kluwer.

Garfield, J., delMas, R., and Zieffler, A. (2010). Developing tertiary-level students' statistical thinking through the use of model-eliciting activities. *Proceedings of the Eighth International Conference on the Teaching of Statistics*. Retrieved on 10.2.2012 from: http://www.stat.auckland.ac.nz/~iase/publications/icots8/ICOTS8_8B3_GARFIELD.pdf

Giaquinto, M. (2007). *Visual thinking in mathematics*. Oxford: Oxford University Press.

Gill, E. and Ben-Zvi, D. (2011). Explanations and context in the emergence of students' informal inferential reasoning. *Mathematical Thinking and Learning*, 13(1 & 2), 87–108.

Gliner, J. A., Leech, N. L., and Morgan, G. A. (2002). Problems with null hypothesis significance testing (NHST): What do the textbooks say? *The Journal of Experimental Education*, 71(1), 83–92.

Godwin, S. and Beswetherick, R. (2003). An investigation into the balance of prescription, experiment and play when learning about the properties of quadratic functions with ICT. *Research in Mathematics Education*, 5(1), 79–95.

Gomez, C., Steinborsdottir, O., and Uselmann, L. (1999). Student's understanding of rate of change: the use of different representation (sic). *Paper presented at the Annual Meeting of the American Educational Research Association*, Montreal, Canada, April 19–23.

Gould, R. (2004). Variability: One statistician's view. *Statistics Education Research Journal*, 3(2), 7–16.

Gravemeijer, K., Bowers, J., and Stephan, M. (2003). A hypothetical learning trajectory on measurement and flexible arithmetic. In M. Stephan, J. Bowers, P. Cobb, and K. Gravemeijer (Eds), *Supporting students' development of measuring conceptions: Analyzing students' learning in social context* (Journal for Research in Mathematics Education Monograph 12) (pp. 51–66). Reston, VA: NCTM.

Green, D. R. (1982). *Probability concepts in 11–16 year old pupils* (2nd edn). Centre for Advancement of Mathematical Education in Technology, University of Technology, Loughborough.

Greeno, J. (1991). Number sense as a situated knowing in a conceptual domain. *Journal for Research in Mathematics Education*, 22(3), 170–218.

Greeno, J. G. (1994). Gibson's affordances. *Psychological Review*, 101(2), 336–342.

Greer, B. and Mukhhopadhyay, S. (2005). Teaching and learning the mathematization of probability: Historical, cultural, social and political contexts. In G. A. Jones (Ed.), *Exploring probability in school: Challenges for teaching and learning* (pp. 297–324). New York: Springer.

Gutiérrez, A., Jaime, A., and Fortuny, J. M. (1991). An alternative paradigm to evaluate the acquisition of the van Hiele levels. *Journal for Research in Mathematics Education*, 22(3), 237–251.

Hadjidemetriou, C. and Williams, J. S. (2010). The linearity prototype for graphs: Cognitive and sociocultural perspectives. *Mathematical Thinking and Learning*, 12(1), 68–85.

Harel, G. and Sowder, L. (2007). Toward a comprehensive perspective on proof. In F. Lester (Ed.), *Second handbook of research on teaching and learning mathematics* (pp. 805–842). Reston, VA: NCTM.

Hart, K. (Ed.) (1981). *Children's understanding of mathematics: 11–16*. London: John Murray.

Hart, K. (1984). Which comes first, length, area, or volume? *Arithmetic Teacher*, 31(9), 16–18 and 26–27.

Hart, K., Brown, M., Kerslake, D., Kuchermann, D., and Ruddock, G. (1985). *Chelsea Diagnostic Mathematics Tests.* Windsor (UK): NFER-Nelson.

Heid, M. K. (1988). Resequencing skills and concepts in applied calculus using the computer as a tool. *Journal for Research in Mathematics Education*, 19(1), 3–25.

Hembree, R. and Dessert, D. (1986). Effects of hand-held calculators in precollege mathematics education: A meta-analysis, *Journal for Research in Mathematics Education*, 17(2), 83–99.

Herbst, P. (2002). Establishing a custom of proving in American school geometry: Evolution of the two-column proof in the early twentieth century. *Educational Studies in Mathematics*, 49(3), 283–312.

Hershkowitz, R. (1989). Visualization in geometry: two sides of the coin. *Focus on Learning Problems in Mathematics*, 11(1 & 2), 61–76.

Hershkowitz, R. (1990). Psychological aspects of learning geometry. In P. Nesher and J. Kilpatrick (Eds), *Mathematics and cognition* (pp. 70–95). Cambridge: Cambridge University Press.

Hershkowitz, R. and Bruckheimer, M. (1981). The quadratic function as a vehicle for discovery by deduction. In C. Comiti and G. Vergnaud (Eds), *Proceedings of the 5th PME International Conference,* Vol. 3 (pp. 193–198). Grenoble, France.

Hestenes, D. (2010). Modeling theory for maths and science education. In R. Lesh, P. Galbraith, C. Haines, and A. Hurford (Eds), *Modeling students' mathematical competencies* (pp. 13–41). New York: Springer.

Hewitt, D. (1996). Mathematical fluency: The nature of practice and the role of subordination. *For the Learning of Mathematics*, 16(2), 28–35.

Hiebert, J. (1981). Units of measure: Results and implications from national assessment. *Arithmetic Teacher*, 28 (6), 38–43.

Hiebert, J. and Wearne, D. (1985). A model of students' decimal computation procedures. *Cognition and Instruction*, 2(3 & 4), 175–205.

Hiebert, J. and Wearne, D. (1986). Procedures over concepts: The acquisition of decimal number knowledge. In J. Hiebert (Ed.), *Conceptual and procedural knowledge: The case of mathematics* (pp. 199–223). Hillsdale, NJ: Erlbaum.

Hino, K. (2002). Acquiring new use of multiplication through classroom teaching: An exploratory study. *Journal of Mathematical Behavior*, 20(1), 477–502.

Hitt, F. (1998). Difficulties in the articulation of different representations linked to the concept of function. *Journal of Mathematical Behavior*, 17(1), 123–134.

Hjalmarson, M. A., Moore, T. J., and delMas, R. (2011). Statistical analysis when the data is an image: Eliciting student thinking about sampling and variability. *Statistics Education Research Journal*, 10(1), 15–34.

Hodgen, J., Küchemann, D., Brown, M., and Coe, R. (2009). School students' understandings of algebra 30 years on. In M. Tzekaki, M. Kaldrimidou, and H. Sakonidis

(Eds), *Proceedings of the 33rd Conference of the International Group for the Psychology of Mathematics Education* (Vol. 3, pp. 177–184). Thessaloniki, Greece.

Hollebrands, K. (2007). The role of a dynamic software program for geometry in the strategies high school mathematics students employ. *Journal for Research in Mathematics Education*, 38(2), 164–192.

Hollebrands, K., Laborde, C., and Strasser, R. (2008). Technology and the learning of geometry at the secondary level. In M. K. Heid and G. Blume (Eds), *Research on technology in the learning and teaching of mathematics, volume 1: Research syntheses* (pp. 155–205). Greenwich, CT: Information Age.

Hölzl, R. (1996). How does 'dragging' affect the learning of geometry. *International Journal of Computers for Mathematical Learning*, 1(2), 169–187.

Horton, R. M., Storm, J., and Leonard, W. H. (2004). The graphing calculator as an aid to teaching algebra. *Contemporary Issues in Technology and Teacher Education*, 4(2), 152–162.

Howden, H. (1989). Teaching number sense. *Arithmetic Teacher*, 36(6), 6–11.

Hoyles, C. and Healy, L. (1997). Unfolding meanings for reflective symmetry. *International Journal of Computers in Mathematical Learning*, 2(1), 27–59.

Hoyles, C. and Jones, K. (1998). Proof in dynamic geometry contexts. In C. Mammana and V. Villani (Eds), *Perspectives on the teaching of geometry for the 21st Century* (pp. 121–128). Dordrecht, The Netherlands: Kluwer.

Hoyles, C., Foxman, D., and Küchemann, D. (2002). *A comparative study of geometry curricula*. London: Qualifications and Curriculum Authority.

Hunter, R. and Anthony, G. (2003). Percentages: A foundation for supporting students' understanding of decimals. In L. Bragg, C. Campbell, G. Herbert, and J. Mousley (Eds), *Mathematics Education Research: Innovation, networking, opportunity* (Proceedings of the 26th annual conference of the Mathematics Education Research Group of Australasia, Freemantle, pp. 452–459). Sydney: MERGA.

Inhelder, B. and Piaget, J. (1958). *Growth of logical thinking from childhood to adolescence*. London: RKP.

Irwin, K. (2001). Using everyday knowledge of decimals to enhance understanding. *Journal for Research in Mathematics Education*, 32(4), 399–422.

Izsak, A. and Findell, B. (2005). Adaptive interpretation: Building continuity between students' experiences solving problems in arithmetic and algebra. *ZDM: The International Journal on Mathematics Education*, 37(1), 60–67.

Jackiw, N. and Sinclair, N. (2009). Sounds and pictures: Dynamism and dualism in dynamic geometry. *ZDM: The International Journal on Mathematics Education*, 41(4), 413–426.

Janvier, C. (1981). Use of situations in mathematics education. *Educational Studies in Mathematics*, 12(1), 113–122.

Janvier, C. (1987). Translation processes in mathematics education. In C. Janvier (Ed.), *Problems of representation in the teaching and learning of mathematics* (pp. 27–32). Hillsdale, NJ: Erlbaum.

JMC (Joint Mathematical Council) (2011). *Digital technologies and mathematics education*. London: JMC.

Johnston-Wilder, S. and Mason, J. (Eds) (2005). *Developing thinking in geometry*. London: Paul Chapman.

Jones, G., Langrall, C. W., and Mooney, E. S. (2007). Research in probability: Responding to classroom realities. In F. K. Jester, Jr. (Ed.), *Second handbook of research on mathematics teaching and learning* (pp. 909–956). Charlotte, NC: Information Age Publishing.

Jones, G., Taylor, A., Gardner, G., and Forrester, J. (2012). Accuracy of measurement estimation: context, units, and logical thinking. *School Science and Mathematics*, 112(3), 171–178.

Jones, I. (2008). A diagrammatic view of the equals sign: Arithmetical equivalence as a means not an end. *Research in Mathematics Education*, 10(2), 119–133.

Jones, I. and Pratt, D. (2006). Connecting the equals sign. *International Journal of Computers for Mathematical Learning*, 11(3), 301–325.

Jones, I. and Pratt, D. (2012). A substituting meaning for the equals sign in arithmetic notating tasks. *Journal for Research in Mathematics Education*, 43(1), 2–33.

Jones, K. (2000a). Critical issues in the design of the geometry curriculum. In Bill Barton (Ed.), *Readings in mathematics education* (pp. 75–90). Auckland, New Zealand: University of Auckland.

Jones, K. (2000b). Providing a foundation for deductive reasoning: Students' interpretations when using dynamic geometry software. *Educational Studies in Mathematics*, 44(1–3), 55–85.

Jones, K. (2001). Spatial thinking and visualisation. In *Teaching and learning geometry 11–19* (pp. 55–56). London: Royal Society.

Jones, K. (2002). Issues in the teaching and learning of geometry. In L. Haggarty (Ed.), *Aspects of teaching secondary mathematics* (pp. 121–139). London: Routledge.

Jones, K. (2005). Using Logo in the teaching and learning of mathematics: A research bibliography. *MicroMath*, 21(3), 34–36.

Jones, K. (2012). Using dynamic geometry software in mathematics teaching: a revised research bibliography. *Mathematics Teaching*, 229, 49–50.

Jones, K. and Mooney, C. (2003). Making space for geometry in primary mathematics. In I. Thompson (Ed.), *Enhancing primary mathematics teaching* (pp. 3–15). London: Open University Press.

Jones, K., Mackrell, K., and Stevenson, I. (2009). Designing digital technologies and learning activities for different geometries. In C. Hoyles and J.-B. Lagrange (Eds),

*Mathematics education and technology: Rethinking the terrain* (ICMI Study 17) (pp. 47–60). New York: Springer.

Joram, E., Gabriele, A., Bertheau, M., *et al.* (2005). Children's use of the reference point strategy for measurement estimation. *Journal for Research in Mathematics Education*, 36(1), 4–23.

Joram, E., Subrahmanyam, K., and Gelman, R. (1998). Measurement estimation: Learning to map the route from number to quantity and back. *Review of Educational Research*, 68(4), 413–449.

Kahneman, D., Slovic, P., and Tversky, A. (1982). *Judgement under uncertainty: Heuristics and biases*. Cambridge: Cambridge University Press.

Kamii, C. and Kysh, J. (2006). The difficulty of "length x width": Is a square the unit of measurement? *Journal of Mathematical Behavior*, 25(2), 105–115.

Kaplan, J. and Du, J. (2009). Question format and representations: Do heuristics and biases apply to statistics students? *Statistics Education Research Journal*, 8(2), 56–73.

Kaput, J. (1989). Linking representations in the symbolic systems of algebra. In S. Wagner and C. Kieran (Eds), *Research agenda for mathematics education: Research issues in the learning and teaching of algebra* (pp. 167–194). Reston, VA: National Council of Teachers of Mathematics.

Kaput, J. (1998). Transforming algebra from an engine of inequity to an engine of mathematical power by "algebrafying" the K–12 curriculum. In National Council of Teachers of Mathematics and Mathematical Sciences Education Board (Eds), *The nature and role of algebra in the K–14 curriculum: Proceedings of a National Symposium* (pp. 25–26). Washington, DC: National Research Council, National Academy Press.

Kaput, J. (1999). Teaching and learning a new algebra. In E. Fennema and T. A. Romberg (Eds), *Mathematics classrooms that promote understanding* (pp. 133–155). Mahwah, NJ: Erlbaum.

Kaput, J. and Maxwell-West, M. (1994). Missing-value proportional reasoning problems: Informal reasoning patterns. In G. Harel and J. Confrey (Eds), *The development of multiplicative reasoning factors affecting in the learning of mathematics* (pp. 235–287). Albany: State University of New York Press.

Kaput, J. and Schorr, R. (2008). Changing representational infrastructures changes most everything: The case of SimCalc, algebra and calculus. In G. W. Blume and M. K. Heid (Eds), *Research on technology and the teaching and learning of mathematics. Cases and perspectives* (pp. 211–253). Cambridge: IAP.

Karplus, E., Karplus, R., and Wollman, W. (1974). Intellectual development beyond elementary school IV: Ratio, the influence of cognitive style. *School Science and Mathematics*, 74, 476–482.

Karplus, R., Pulos, S., and Stage, E. K. (1983). Early adolescents' proportional reasoning on 'rate' problems. *Educational Studies in Mathematics*, 14(3), 219–233.

Keiser, J. M. (2004). Struggles with developing the concept of angles: Comparing sixth-grade students' discourse to the history of the angle concept. *Mathematical Thinking and Learning*, 6(3), 285–306.

Kelly, D. L., Mullis, I. V. S., and Martin, M. O. (2000). *Profiles of student achievement in mathematics at the TIMSS international benchmarks: U.S. performance and standards in an international context*. Chestnut Hill, MA: Boston College.

Kemp, A. (2009). Trigonometry from first principles. *Mathematics Teaching*, 215, 40–41.

Kendal, M. and Stacey, K. (1996). Trigonometry: Comparing ratio and unit circle methods. In P. Clarkson (Ed.), *Technology in Mathematics Education: Proceedings of the 19th Annual Conference of the Mathematics education Research group of Australasia* (pp. 322–329). Melbourne: MERGA.

Kent, P., Bakker, A., Hoyles, C., and Noss, R. (2011). Measurement in the workplace: The case of process improvement in manufacturing industry. *ZDM: The International Journal on Mathematics Education*, 43(5), 747–758.

Kidman, G. C. (1999). Grade 4, 6 and 8 students' strategies in area measurement. In J. M. Truran and K. M. Truran (Eds), *Making the Difference* (Proceedings of the 22nd Annual Conference of the Mathematics Education Research Group of Australasia, pp. 298–305). MERGA, Adelaide.

Kieran, C. (1981). Concepts associated with the equality symbol. *Educational Studies in Mathematics*, 12(3), 317–326.

Kieran, C. (1984). A comparison between novice and more-expert algebra students on tasks dealing with the equivalence of equations. In J. M. Moser (Ed.), *Proceedings of the Sixth Annual Meeting of the Group for the Psychology of Mathematics Education in North America*. Madison, Wisconsin.

Kieran, C. (1986). LOGO and the notion of angle among fourth and sixth grade children. In *Proceedings of the 10th Conference of the International Group for the Psychology of Mathematics Education* (Vol. 1, pp. 99–104), London, England.

Kieran, C. (1992). The learning and teaching of algebra. In D.A. Grouws (Ed.), *Handbook of Research on Mathematics Teaching and Learning* (pp. 390–419). New York: Macmillan.

Kieren, T. and Southwell, B. (1979). Rational numbers as operators: The development of this construct in children and adolescents. *Alberta Journal of Educational Research*, 25(4), 234–247.

Kloosterman, P., Warfield, J., Wearne, D., *et al.* (2004). Knowledge of learning mathematics and perceptions of learning mathematics in 4th-grade students. In P. Kloosterman and F. Lester (Eds), *Results and interpretations of the 2003 mathematics assessment of the NAEP*. Reston, VA: NCTM.

Konold, C. (1989). Informal conceptions of probability. *Cognition and Instruction*, 6, 59–98.

Konold, C. (1991). Understanding students' beliefs about probability. In E. von Glasersfeld (Ed.), *Radical constructivism in mathematics education* (pp. 139–156), Amsterdam, The Netherlands: Kluwer.

Konold, C. (1995). Issues in assessing conceptual understanding in probability and statistics. *Journal of Statistics Education*, 3(1). Retrieved on 22.10.2010 from: http://www.amstat.org/publications/jse/v3n1/konold.html [accessed 4 August 2004].

Konold, C. and Pollatsek, A. (2002). Data analysis as the search for signals in noisy processes. *Journal for Research in Mathematics Education*, 33(4), 259–289.

Konold, C. and Higgins, T. L. (2003). Reasoning about data. In J. Kilpatrick, W. G. Martin, and D. Schifter (Eds), *A research companion to Principles and Standards for School Mathematics*, (pp. 193–215). Reston, VA: National Council of Teachers of Mathematics.

Konold, C., Pollatsek, A., Well, A. D., *et al.* (1993). Inconsistencies in students' reasoning about probability. *Journal for Research in Mathematics Education*, 24(5), 392–414.

Konold, C., Harradine, A., and Kazak, S. (2007). Understanding distributions by modeling them. *International Journal of Computers for Mathematical Learning*, 12(3), 217–230.

Küchemann, D. (1981). Algebra. In K. Hart (Ed.), *Children's understanding of mathematics 11–16* (pp. 102–119). London: Murray.

Kuntze, S., Lerman, S., Murphy, B., *et al.* (2011). *Awareness of Big Ideas in mathematics classrooms: Final report.* Audiovisual and Culture Executive Agency, European Commission. Pädagogische Hochschule, Ludwigsburg.

Kurtz, B. and Karplus, R. (1979). Intellectual development beyond elementary school VII: Teaching for proportional reasoning. *School Science and Mathematics*, 79, 387–398.

Laborde, C. (1993). The computer as part of the learning environment: The case of geometry. In C. Keitel and K. Ruthven (Eds), *Learning through computers: Mathematics and educational technology* (pp. 48–67). Berlin, Germany: Springer.

Laborde, C. (1998). Relationships between the spatial and theoretical in geometry: The role of computer dynamic representations in problem solving. In J. Tinsley and D. Johnson (Eds), *Information and communication technologies in school mathematics* (pp. 183–194). London: Chapman Hall.

Laborde, C. (2001). Integration of technology in the design of geometry tasks with Cabri-Geometry. *International Journal of Computers for Mathematical Learning*, 6(3), 283–317.

Laborde, C. (2004). The hidden role of diagrams in pupils' construction of meaning in geometry. In J. Kilpatrick, C. Hoyles, and O. Skovsmose (Eds), *Meaning in mathematics education* (pp. 159–179). Dordrecht, The Netherlands: Kluwer.

Laborde, C., Kynigos, C., Hollebrands, K., and Strasser, R. (2006). Teaching and learning geometry with technology. In A. Gutiérrez and P. Boero (Eds), *Handbook of research on the psychology of mathematics education: Past, present and future* (pp. 275–304). Rotterdam: Sense Publishers.

Lachance, A. and Confrey, J. (2002). Helping students build a path of understanding from ratio and proportion to decimal notation. *Journal of Mathematical Behavior*, 20(4), 503–526.

Lakatos, I. (1976). *Proofs and refutations.* Cambridge: Cambridge University Press.

Lakoff, G. and Núñez, R. (2000). *Where mathematics comes from: How the embodied mind brings mathematics into being.* New York: Basic Books.

Lamon, S. J. (1993). Ratio and proportion: Connecting content and children's thinking. *Journal for Research in Mathematics Education*, 24(1), 41–61.

Lamon, S. J. (1996). The development of unitizing: Its role in children's partitioning strategies. *Journal for Research in Mathematics Education*, 27(2), 170–193.

Langrall, C., Nisbet, S., Mooney, E., and Jansem, S. (2011). The role of context expertise when comparing data. *Mathematical Thinking and Learning*, 13(1 & 2), 47–67.

Lecoutre, M. P. (1992). Cognitive models and problem spaces in "purely random" situations. *Educational Studies in Mathematics*, 23(6), 589–593.

Lee, L. (1996) An initiation into algebraic culture through generalization activities. In N. Bednarz, C. Kieran, and L. Lee (Eds), *Approaches to algebra: Perspectives for research on teaching* (pp. 87–106). Dordrecht, The Netherlands: Kluwer.

Lehrer, R. (2003). Developing understanding of measurement. In J. Kilpatrick, W. G. Martin and D. Schifter (Eds), *A research companion to Principles and Standards for School Mathematics* (pp. 179–192). Reston, VA: NCTM.

Lehrer, R., Jenkins, M., and Osana, H. (1998). Longitudinal study of children's reasoning about space and geometry. In R. Lehrer and D. Chazan (Eds), *Designing learning environments for developing understanding of geometry and space* (pp. 137–167). Mahwah, NJ: Erlbaum.

Lehrer, R., Jacobson, C., Kemeny, V., and Strom, D. (1999). Building on children's intuitions to develop mathematical understanding of space. In E. Fennema and T. A. Romberg (Eds), *Mathematics classrooms that promote understanding* (pp. 63–87). Mahwah, NJ: Erlbaum.

Lehrer, R., Jaslow, L., and Curtis, C. (2003). Developing an understanding of measurement in the elementary grades. In D. H. Clements and G. Bright (Eds), *Learning and teaching measurement* (pp. 81–99). Reston, VA: National Council of Teachers of Mathematics.

Leikin, R., Berman, A., and Zaslavsky, O. (2000). Learning through teaching: The case of symmetry. *Mathematics Education Research Journal*, 12(1), 16–34.

Leinhardt, G., Zaslavsky, O., and Stein, M. S. (1990). Functions, graphs and graphing: Tasks, learning, and teaching. *Review of Educational Research*, 1(1), 1–64.

Leung, A. (2011). An epistemic model of task design in dynamic geometry environment. *ZDM: The International Journal on Mathematics Education*, 43(3), 325–336.

Levenson, E., Tirosh, D., and Tsamir, P. (2011). *Preschool geometry: Theory, research, and practical perspectives*. Rotterdam, The Netherlands: Sense Publishers.

Lima, R. and Tall, D. (2008). Adaptive interpretation: Building continuity between students' experiences solving problems in procedural embodiment and magic in linear equations. *Educational Studies in Mathematics*, 67(1), 3–18.

Linchevski, L. and Herscovics, N. (1996). Crossing the cognitive gap between arithmetic and algebra: Operating on the unknown in the context of equations. *Educational Studies in Mathematics*, 30(1), 39–65.

Lineberry, C. A. and Keene, K. A. (2011). Using dynamic geometry to develop students' conceptual understanding of angle. In L. R. Wiest and T. Lamberg (Eds), *Proceedings of the 33rd annual meeting of the North American chapter of the international group for the psychology of mathematics education* (pp. 1024–1032). Reno, NV: University of Nevada, Reno.

Lobato, J. and Thanheiser, E. (2002). Developing understanding of ratio-as-measure as a foundation for slope. In B. Litwiller and G. Bright (Eds), *Making sense of fractions, ratios, and proportions* (pp. 162–175). Reston, VA: National Council of Teachers of Mathematics.

MAA and NCTM (2012). *Calculus: A joint position statement of the Mathematical Association of America and the National Council of Teachers of Mathematics*. Reston, VA: NCTM.

MacGregor, M. and Stacey, K. (1993). Cognitive models underlying students' formulation of simple linear equations. *Journal for Research in Mathematics Education*, 24(3), 217–232.

MacGregor, M. and Stacey, K. (1997). Students' understanding of algebraic notation 11–15. *Educational Studies in Mathematics*, 33(1), 1–19.

Mackrell, K. (2011). Design decisions in interactive geometry software. *ZDM: The International Journal on Mathematics Education*, 43(3), 373–387.

Makar, K. and Rubin, A. (2007). Beyond the bar graph: Teaching statistical inference in primary school. *The Fifth International Forum for Research on Statistical Reasoning, Thinking and Literacy*, The University of Warwick, UK.

Makar, K. and Rubin, A. (2009). A framework for thinking about informal statistical inference. *Statistics Education Research Journal*, 8(1), 82–105.

Makar, K., Bakker, A., and Ben-Zvi, D. (2011). The reasoning behind informal statistical inference. *Mathematical Thinking and Learning*, 13(1 & 2), 152–173.

Malkevitch, J. (2009). What is geometry? In T. Craine and R. Rubenstein (Eds), *Understanding geometry for a changing world*, 71st Yearbook (pp. 3–16). Reston, VA: NCTM.

Malkevitch, J. (Ed.) (1992). *Geometry's future*. Consortium for Mathematics and Its Applications, Lexington, MA. 2nd edition.

Martinez, M. and Brizuela, B. (2006). A third grader's way of thinking about linear function tables. *Journal of Mathematical Behavior*, 25(4), 285–298.

Martinez-Sierra, G. (2008). From the analysis of the articulation of the trigonometric functions to the corpus of Eulerian analysis to the interpretation of the conceptual breaks present in its scholar structure, *Proceedings of the 2008 History and Pedagogy of Mathematics conference*. Mexico City, Mexico.

Mason, J. and Sutherland, R. (2002). *Key aspects of teaching algebra in schools*. London: QCA.

Maymon-Erez, M. and Yerushalmy, M. (2007). "If you can turn a rectangle to a square then you can turn a square to a rectangle...": On the complexity and importance of psychologizing the dragging tool by young students. *International Journal of Computers for Mathematical Learning*, 11(3), 271–299.

McCrone, S. M. S. and Martin, T. S. (2009). Formal proof in high school geometry: Student perceptions of structure, validity, and purpose. In D. A. Stylianou, M. L. Blanton, and E. J. Knuth (Eds), *Teaching and learning proof across the grades* (pp. 204–221). New York: Routledge.

Meserve, B. E. (1967). Euclidean and other geometries. *Mathematics Teacher*, 60(1), 2–11.

Mevarech, Z. and Kramarsky, B. (1997). From verbal descriptions to graphic representations: stability and change in students' alternative conceptions. *Educational Studies in Mathematics*, 32(3), 229–263.

Michelsen, C. (2006). Functions: A modelling tool in mathematics and science. *ZDM: The International Journal on Mathematics Education*, 38(3), 269–280.

Miller, T. K. (1998). The random variable concept in introductory statistics. *Proceedings of the fifth international conference on the teaching of statistics*. Retrieved on 10.2.2012 from: http://www.stat.auckland.ac.nz/~iase/publications/2/Topic9h.pdf [accessed 4 August 2012].

Mitchelmore, M. C. and White, P. (1998). Development of angle concepts: A framework for research. *Mathematics Education Research Journal*, 10(3), 4–27.

Mitchelmore, M., White, P., and McMaster, H. (2007). Teaching ratio and rates for abstraction. In J. Watson and K. Beswick (Eds), *Proceedings of the 30th Annual Conference of the Mathematics Education Research Group of Australasia*, Hobart, Australia, MERGA.

Modestou, M. and Gagatsis, A. (2010). Cognitive and metacognitive aspects of proportional reasoning. *Mathematical Thinking and Learning*, 12(1), 36–53.

Moschkovich, J. (1998). Students' use of the x-intercept as an instance of a transitional conception. *Educational Studies in Mathematics*, 37(2), 169–197.

Moss, J. and Case, R. (1999). Developing children's understanding of rational numbers: A new model and an experimental curriculum. *Journal for Research in Mathematics Education*, 30(2), 122–147.

Mullis, I., Martin, M., and Foy, P. (2008). *TIMSS 2007 international mathematics report: Findings from IEA's Trends in International Mathematics and Science Study at the fourth and eighth Grades*. Boston MA: TIMSS and PIRLS International Study Center, Boston College.

NCTM (National Council of Teachers of Mathematics) (2009). *Focus in high school mathematics: Reasoning and sense making*. Reston, VA: National Council of Teachers of Mathematics.

Nelsen, R. B. (1993). *Proofs without words: Exercises in visual thinking*. Washington, DC: Mathematical Association of America.

Newcombe, N. S., Uttal, D. H., and Sauter, M. (2012). Spatial development. In P. Zelazo (Ed.), *Oxford Handbook of Developmental Psychology*. Oxford: Oxford University Press.

Nissen, P. (2000). A geometry solution from multiple perspectives. *Mathematics Teacher*, 93(4), 324–327.

NMAP, National Mathematics Advisory Panel (2008). Foundations for success: The final report of the National Mathematics Advisory Panel. Retrieved on 2.4.2008 from: http://www.ed.gov/about/bdscomm/list/mathpanel/index.html [accessed 4 August 2012].

Noelting, G. (1980a). The development of proportional reasoning and the ratio concept Part I – Differentiation of stages. *Educational Studies in Mathematics*, 11(2), 217–253.

Noelting, G. (1980b). The development of proportional reasoning and the ratio concept Part II – Problem-structure at successive stages: Problem-solving strategies and the mechanism of adaptive restructuring. *Educational Studies in Mathematics*, 11(3), 331–363.

Noll, J. (2011). Graduate teaching assistants' statistical content knowledge of sampling. *Statistics Education Research Journal*, 10(2), 48–74.

Noll, J., Shaughnessy, M., and Ciancetta, M. (2010). Students' statistical reasoning about distribution across grade levels: A look from middle school through graduate school. *Proceedings of the Eighth International Conference on the Teaching of Statistics*. Retrieved on 10.2.2012 from: http://www.stat.auckland.ac.nz/~iase/publications/icots8/ICOTS8_8B4_NOLL.pdf [accessed 4 August 2012].

Nunes, T. and Bryant, P. (2009). Paper 3 Understanding rational numbers and intensive quantities. In T. Nunes, P. Bryant and A. Watson, *Key understandings in mathematics learning: A report to the Nuffield Foundation*. Retrieved on 22.10.2010 from: http://www.nuffieldfoundation.org/key-understandings-mathematics-learning [accessed 4 August 2012].

Nunes, T., Schliemann, A. D., and Carraher, D. W. (1993). *Street mathematics and school mathematics*. Cambridge: Cambridge University Press.

Nunes, T., Desli, D., and Bell, D. (2003). The development of children's understanding of intensive quantities. *International Journal of Educational Research*, 39, 652–675.

Nunes, T., Bryant, P., and Watson, A. (2009). *Key understandings in mathematics learning*. London: Nuffield Foundation. Retrieved on 22.10.2010 from: http://www.nuffieldfoundation.org/key-understandings-mathematics-learning [accessed 4 August 2012].

O'Brien, T. C. (1968). An experimental investigation of a new approach to the teaching of decimals [PhD Thesis, New York University, 1967], *Dissertation Abstracts*, 28(11), 4541.

O'Callaghan, B. R. (1998). Computer-intensive algebra and students' conceptual knowl-edge of functions. *Journal for Research in Mathematics Education*, 29(1), 21–40.

Oldknow, A. and Taylor, R. (Eds) (1998). *Data-capture and modelling in mathematics and science*. Coventry: BECTa.

Olivero, F. and Robutti, O. (2007). Measuring in dynamic geometry environments as a tool for conjecturing and proving. *International Journal of Computers for Mathematical Learning*, 12(2), 135–156.

Oner, D. (2008). A comparative analysis of high school geometry curricula: What do technology-intensive, standards-based, and traditional curricula have to offer in terms mathematical proof and reasoning? *Journal of Computers in Mathematics and Science Teaching*, 27(4), 467–497.

Orhun, N. (2001). Student's mistakes and misconceptions on teaching of trigonometry. *Mathematics Education into the 21st Century Project: Proceedings of the International Conference on New Ideas in Mathematics Education*, 19–24 August 2001, Australia.

Orton, A. (1983a). Students' understanding of integration. *Educational Studies in Mathematics*, 14(1), 1–18.

Orton, A. (1983b). Student's understanding of differentiation. *Educational Studies in Mathematics*, 14(3), 235–250.

Outhred, L. and Mitchelmore, M. (2000). Young children's intuitive understanding of rectangular area measurement. *Journal for Research in Mathematics Education*, 31(2), 144–167.

Owens, K. and Outhred, L. (2006). The complexity of learning geometry and measure-ment. In A. Gutiérrez and P. Boero (Eds), *Handbook of research on the psychology of mathematics education: Past, present and future* (pp. 83–115). Rotterdam, The Netherlands: Sense Publishers.

Parzysz, B. (1988). Knowing vs. Seeing: Problems of the plane representation of space geometry figures. *Educational Studies in Mathematics*, 19(1), 79–92.

Patronis, T. (1994). On students' conception of axioms in school geometry. *Proceedings of the 18th Conference of the International Group for the Psychology of Mathematics Education* (Vol. 4, pp. 9–16). Lisbon, Portugal.

Perks, P. and Prestage, S. (2006). The ubiquitous isosceles triangle part 1: Constructions. *Mathematics in School*, 35(1), 2–3.

Perrenet, J. and Wolters, M. (1994). The art of checking: A case study of students' errone-ous checking behavior in introductory algebra. *Journal of Mathematical Behavior*, 13(3), 335–358.

Pfannkuch, M. (2005). Thinking tools and variation. *Statistics Education Research Journal*, 4(1), 83–91.

Pfannkuch, M. (2011). The role of context in developing informal statistical inferential reasoning: A classroom study. *Mathematical Thinking and Learning*, 13(1 & 2), 27–46.

Phillips, L. M., Norris, S. P., and Macnab, J. S. (2010). *Visualization in mathematics, reading and science education*. New York: Springer.

Piaget, J. and Inhelder, B. (1948/1956). *The child's conception of space*. London: Routledge (original edition in French, 1948).

Piaget, J. (1953). How children form mathematical concepts. *Scientific American*, 189(5), 74–79.

Piaget, J. and Inhelder, B. (1975). *The origin of the idea of chance in children* (L. J. Leake and P. D. Burell and H. D. Fischbein, Trans.). London: Routledge & Kegan Paul. (Original work published in 1951.)

Piaget, J., Inhelder, B., and Szeminska, A. (1948/1960). *The child's conception of geometry* (E. A. Lunzer, Trans.). New York: Basic Books (original edition in French, 1948).

Pierce, R. and Stacey, K. (2011). Using dynamic geometry to bring the real world into the classroom. In L. Bu and R. Schoen (Eds.), *Model-centered learning: Pathways to mathematical understanding using GeoGebra* (pp. 41–55). Rotterdam: Sense Publishers.

Pirie, S. and Martin, L. (1997). The equation, the whole equation and nothing but the equation! One approach to the teaching of linear equations. *Educational Studies in Mathematics*, 34(2), 159–181.

Pizlo, Z. (2008). *3D Shape: Its unique place in visual perception*. Cambridge, MA: MIT Press.

Pollatsek, A., Lima, S., and Well, A. D. (1981). Concept or computation: Students' understanding of the mean. *Educational Studies in Mathematics*, 12(2), 191–204.

Pratt, D. (2000). Making sense of the total of two dice. *Journal for Research in Mathematics Education*, 31(5), 602–625.

Pratt, D. and Noss, R. (2002). The micro-evolution of mathematical knowledge: The case of randomness. *Journal of the Learning Sciences*, 11(4), 453–488.

Pratt, D., Johnston-Wilder, P., Ainley, J., and Mason, J. (2008). Local and global thinking in statistical inference. *Statistics Education Research Journal*, 7(2), 107–129.

Prediger, S. (2008). Do you want me to do it with probability or with my normal thinking? Horizontal and vertical views on the formation of stochastic conceptions. *International Electronic Journal of Mathematics Education*, 3(3), 126–154.

Presmeg, N. C. (1986). Visualization in high school mathematics. *For the Learning of Mathematics*, 6(3), 42–46.

Prestage, S. and Perks, P. (2001). *Adapting and extending tasks for the secondary classroom: New tasks for old*. London: David Fulton.

Price, M. (2003). A century of school geometry. In C. Pritchard (Ed.), *The changing shape of geometry: Celebrating a century of geometry and geometry teaching*. Cambridge: Cambridge University Press.

Pritchard, L. and Simpson, A. (1999). The role of pictorial images in trigonometry problems. In O. Zaslavsky (Ed.), *Proceedings of the 23rd Conference of the*

*International Group for the Psychology of Mathematics Education* (Vol. 4, pp. 81–88). Haifa, Israel.

Prodromou, T. and Pratt, D. (2006). The role of causality in the coordination of two perspectives on distribution within a virtual simulation. *Statistics Education Research Journal,* 5(2), 69–88.

Psycharis, G. and Kynigos, C. (2004). Normalising geometrical constructions: A context for the generation of meanings for ratio and proportion. *Proceedings of the 28th Conference of the International Group for the Psychology of Mathematics Education* (Vol. 4, pp. 65–72). Bergen, Norway.

Quesada, A., Einsporn, R., and Wiggins, M. (2008). The impact of the graphical approach on students' understanding of the formal definition of limit. *The International Journal for Technology in Mathematics Education,* 15(3), 95–102.

Radford, L. (2008). Iconicity and contraction: A semiotic investigation of forms of algebraic generalizations of patterns in different contexts. *ZDM: The International Journal on Mathematics Education,* 40(2), 83–96.

Ramsey, J. B. (1999). Why do students find statistics so difficult? *Proceedings of the 52nd Session of the International Statistics Institute.* Retrieved on 10.2.2012 from: http://www.stat.auckland.ac.nz/~iase/publications/5/rams0070.pdf.

Robert, A. and Speer, N. (2001). Research on the teaching and learning of calculus/elementary analysis. In D. Holton (Ed.), *The teaching and learning of mathematics at university level: An ICMI study* (pp. 283–299). Dordrecht, The Netherlands: Kluwer.

Roche, A. and Clarke, D. M. (2004). When does successful comparison of decimals reflect conceptual understanding? In I. Putt, R. Farragher, and M. McLean (Eds), *Mathematics Education for the Third Millennium: Towards 2010* (Proceedings of the 27th annual conference of the Mathematics Education Research Group of Australasia) (pp. 486–493). Sydney: MERGA.

Ronda, E. (2009). Growth points in students' developing understanding of function in equation form. *Mathematics Education Research Journal,* 21(1), 31–53.

Roschelle, J., Kaput, J. J., and Stroup, W. (2000). SimCalc: Accelerating student engagement with the mathematics of change. In M. J. Jacobsen and R. B. Kozma (Eds), *Innovations in Science and Mathematics Education: Advanced designs for technologies of learning* (pp. 47–75). Hillsdale, NJ: Erlbaum.

Roschelle, J., Schechtman, N., Tatar, D., *et al.* (2010). Integration of technology, curriculum, and professional development for advancing middle school mathematics: three large-scale studies. *American Educational Research Journal,* 47(4), 833–878.

Roseman, D. (1998). Exploring low dimensional objects in high dimensional spaces. In H.-C. Hege and K. Polthier (Eds), *Mathematical visualization: Algorithms, applications and numerics* (pp. 281–291). Berlin: Springer.

Roth, W.-M. (2011). *Geometry as objective science in elementary classrooms: Mathematics in the flesh.* New York: Routledge.

Royal Society (2001). *Teaching and learning geometry 11–19.* London: Royal Society.

Ruthven, K. (1990). The influence of graphic calculator use on translation from graphic to symbolic forms. *Educational Studies in Mathematics*, 21(5), 431–450.

Ruthven, K., Hennessy, S., and Deaney, R. (2005). Current practice in using dynamic geometry to teach about angle properties. *Micromath*, 21(1), 9–13.

Sajka, M. (2003) A secondary school student's understanding of the concept of function – A case study. *Educational Studies in Mathematics*, 53(3), 229–254.

Saldanha, L. A. and Thompson, P. W. (2002). Conceptions of sample and their relationship to statistical inference. *Educational Studies in Mathematics*, 51(3), 257–270.

Sarama, J., Clements, D. H., Barrett, J., *et al.* (2011). Evaluation of a learning trajectory for length in the early years. *ZDM: The International Journal on Mathematics Education*, 43(5), 667–680.

Schmidt, W. H., McKnight, C., Valverde, G. A., *et al.* (1997). *Many visions, many aims: A cross-national investigation of curricular intentions in school mathematics.* Dordrecht, The Netherlands: Kluwer.

Schmittau, J. (2005). The development of algebraic thinking: A Vygotskian perspective. *ZDM: The International Journal on Mathematics Education*, 37(1), 16–22.

Schuster, S. (1971). On the teaching of geometry. *Educational Studies in Mathematics*, 4(1), 76–86.

Schwarz, B. and Hershkowitz, R. (1999). Prototypes: Brakes or levers in learning the function concept? The role of computer tools. *Journal for Research in Mathematics Education*, 30(4), 362–389.

Schweiger, F. (2006). Fundamental ideas: A bridge between mathematics and mathematical education. In J. Maass and W. Schlöglmann (Eds), *New mathematics education research and practice* (pp. 63–73). Rotterdam: Sense Publishers.

Senk, S. L. and Hirshhorn, D. B. (1990). Multiple approaches to geometry: Teaching similarity. *Mathematics Teacher*, 83(4), 274–280.

Sfard, A. and Linchevski, L. (1994). The gains and the pitfalls of reification: The case of algebra. *Educational Studies in Mathematics*, 26(2–3), 191–228.

Shaughnessy, J. M. (1997). Missed opportunities in research on the teaching and learning of data and chance. In F. Biddulph and K. Carr (Eds), *Proceedings of the Twentieth Annual meeting of the Mathematics Education Research Group of Australasia* (pp. 6–22). Aotearoa, NZ.

Shaughnessy, J. M. (2007). Research on statistics learning and reasoning. In F. K. Lester (Ed.), *Second handbook of research on mathematics teaching and learning* (pp. 957–1009). Charlotte, NC: Information Age Publishing.

Shaughnessy, J. M. and Ciancetta, M. (2002). Students' understanding of variability in a probability environment. In B. Philips (Ed.), *Proceedings of the Sixth International Conference on Teaching Statistics* (CD-ROM), South Africa. The Netherlands: ISI.

Shaughnessy, J. M., Garfield, J., and Greer, B. (1996). Data handling. In A. J. Bishop, K. Clements, C. Keital, *et al.* (Eds), *International handbook of mathematics education* (pp. 205–237). Dordrecht, The Netherlands: Kluwer.

Siemon, D. (2011). Realising the big ideas in number – Vision impossible? *Curriculum Perspectives*, 31(1), 66–69.

Sinclair, M. (2003). Some implications of the results of a case study for the design of pre-constructed dynamic geometry sketches and accompanying materials. *Educational Studies in Mathematics*, 52(3), 289–317.

Sinclair, N. (2008). *The history of the geometry curriculum in the United States.* Charlotte, NC: Information Age Publishing.

Slavit, D. (1997). An alternative route to reification of a function. *Educational Studies in Mathematics*, 33(3), 259–281.

Smith, A. (2004). *Making mathematics count: The report of Professor Adrian Smith's inquiry into post-14 mathematics education.* Norwich: The Stationery Office Ltd.

Smith, J. P., diSessa, A., and Roschelle, J. (1993). Misconceptions reconceived: A constructivist analysis of knowledge in transition. *Journal of the Learning Sciences*, 3(2), 115–163.

Sotos, A. E. C., Vanhoof, S., Van Den Noortgate, W., and Onghena, P. (2009). The transitivity misconception of Pearson's Correlation Coefficient. *Statistics Education Research Journal*, 8(2), 33–55.

Sowder, J., Wearne, D., Martin, G., and Strutchens, M. (2004). What grade 8 students know about mathematics: Changes over a decade. In P. Kloosterman and F. Lester (Eds), *The 1990 through 2000 mathematics assessment of the NAEP: Results and interpretations* (pp. 105–143). Reston, VA: NCTM.

Spyrou, P. and Zagorianakos, A. (2010). Greek students' understandings of the distinction between function and relation. *Research in Mathematics Education*, 12(2), 163–164.

Stacey, K. (1989). Finding and using patterns in linear generalising problems. *Educational Studies in Mathematics*, 20(2), 147–164.

Stacey, K. and MacGregor, M. (1999). Learning the algebraic methods of solving problems. *Journal of Mathematical Behavior*, 18(2), 149–167.

Steer, J., de Vila, M., and Eaton, J. (2009a). Trigonometry with Year 8: Part 1. *Mathematics Teaching*, 214, 42–45.

Steer, J., de Vila, M., and Eaton, J. (2009b). Trigonometry with Year 8: Part 2. *Mathematics Teaching i*, 214i. Retrieved on 22.10.2010 from: http://www.atm.org.uk/journal/archive/mt214ifiles/trigonometry-year8.html [accessed 4 August 2012].

Steer, J., de Vila, M., and Eaton, J. (2009c). Trigonometry with Year 8: Part 3. *Mathematics Teaching*, 215, 6–8.

Steinberg, R. M., Sleeman, D. H., and Ktorza, D. (1990). Algebra students' knowledge of equivalence of equations. *Journal for Research in Mathematics Education,* 22(2), 112–121.

Steinle, V. and Stacey, K. (2003). Grade-related trends in the prevalence and persistence of decimal misconceptions. *Proceedings of the 27th Conference of the International Group for the Psychology of Mathematics Education* (Vol. 4, pp. 259–266). Hawaii, USA.

Steinle, V. and Stacey, K. (2004). A longitudinal study of students' understanding of decimal notation: An overview and refined results. In I. Putt, R. Faragher, and M. McLean (Eds), *Mathematics education for the third millennium: Towards 2010, Proceedings of the 27th annual conference of the mathematics education research group of Australasia* (Vol. 2, pp. 541–48). Townsville: MERGA.

Stevens, S. S. (1946). On the theory of scales of measurement. *Science*, 103(2684), 677–680.

Streefland, L. (1984). Search for the roots of ratio: Some thoughts on the long term learning process (Towards … a theory): Part I: Reflections on a teaching experiment. *Educational Studies in Mathematics*, 15(4), 327–348.

Stroup, W. (2003). Understanding qualitative calculus: A structural synthesis of learning research. *International Journal of Computers for Mathematical Learning*, 7(2), 167–215.

Sutherland, R. and Rojano, T. (1993). A spreadsheet approach to solving algebra problems. *Journal of Mathematical Behavior*, 12(4), 351–383.

Swan, M. (1980). *The language of functions and graphs.* Nottingham, UK: Shell Centre for Mathematical Education. University of Nottingham.

Swinyard, C. (2011). Reinventing the formal definition of limit: The case of Amy and Mike. *Journal of Mathematical Behavior*, 7(4), 765–790.

Tall, D. (1985). Understanding the calculus. *Mathematics Teaching*, 110, 49–53.

Tall, D. (Ed.) (1991). *Advanced mathematical thinking.* Dordrecht, The Netherlands: Kluwer.

Tall, D. (1992). The transition to advanced mathematical thinking: Functions, limits, infinity, and proof. In D. A. Grouws (Ed.), *Handbook of research on mathematics teaching and learning* (pp. 495–511). New York: Macmillan.

Tall, D. (1993). Students' difficulties in calculus. *Proceeding of Working Group 3 on Students' Difficulties in Calculus* (pp. 13–28). Québec, Canada: ICME-7.

Tall, D. (1996). Functions and calculus. In A. Bishop, K. Clements, C. Keital, J. Kilpatrick, and C. Laborde (Eds), *International handbook of mathematics education* (pp. 289–325). Dordrecht, The Netherlands: Kluwer.

Tall, D. (2008). The transition to formal thinking in mathematics. *Mathematics Education Research Journal*, 20(2), 5–24.

Tall, D., Smith, D., and Piez, C. (2008). Technology and calculus. In M. K. Heid and G. M. Blume (Eds), *Research on technology and the teaching and learning of mathematics, Volume I: Research syntheses* (pp. 207–258). Charlotte, NC: Information Age Publishing.

Teppo, A. and Esty, W. (1994). Problem solving using arithmetic and algebraic thinking. *Paper presented at the annual conference of the North American chapter of the international group for the psychology of mathematics education.* Baton Rouge, Louisiana.

The Chicago School Mathematics Project staff (1971). The CSMP development in geometry. *Educational Studies in Mathematics,* 3(3 & 4), 281–285.

Thomas, M. and Tall, D. (2001). The long-term cognitive development of symbolic algebra. In H. Chick, K. Stacey, J. Vincent, and J. Vincent (Eds), *Proceedings of the 12th ICMI study conference: The future of the teaching and learning of algebra* (pp. 590–597). University of Melbourne, Australia.

Thomas, M., Wilson, A., Corballis, M., *et al.* (2010). Evidence from cognitive neuroscience for the role of graphical and algebraic representations in understanding function. *ZDM: The International Journal on Mathematics Education,* 42(6), 607–619.

Thompson, P. W. (1994). Images of rate and operational understanding of the fundamental theorem of calculus. *Educational Studies in Mathematics,* 26(2–3), 229–274.

Thompson, P. W. and Saldanha, L. A. (2003). Fractions and multiplicative reasoning. In J. Kilpatrick, W. Gary Martin, and D. Schifter (Eds), *A research companion to the Principles and Standards for School Mathematics* (pp. 95–113). Reston, VA: The National Council of Teachers of Mathematics.

Thompson, P. W. and Silverman, J. (2008). The concept of accumulation in calculus. In M. P. Carlson and C. Rasmussen (Eds), *Making the connection: Research and teaching in undergraduate mathematics* (pp. 43–52). Washington, DC: Mathematical Association of America.

Threlfall, J. (2004). Uncertainty in mathematics teaching: the National Curriculum experiment in teaching probability to primary pupils. *Cambridge Journal of Education,* 34(3), 297–314.

Tirosh, D., Even, R., and Robinson, M. (1998). Simplifying algebraic expressions: Teacher awareness and teaching approaches. *Educational Studies in Mathematics,* 35(1), 51–64.

Toluk, Z. and Middleton, J. (2003). The development of children's understanding of the quotient: A teaching experiment. Retrieved on 22.10.2010 from: http://www.cimt.plymouth.ac.uk/journal/middleton.pdf [accessed 4 August 2012].

Tourniaire, F. and Pulos, S. (1985). Proportional reasoning: A review of the literature. *Educational Studies in Mathematics,* 16(2), 181–204.

Trgalová, J., Soury-Lavergne, S., and Jahn, A. P. (2011). Quality assessment process for dynamic geometry resources in Intergeo project: rationale and experiments. *ZDM: The International Journal on Mathematics Education,* 43(3), 337–351.

Tukey, J. (1977). *Exploratory data analysis.* Reading, MA: Addison-Wesley.

Turnau, S. (2002). Geometry through open problems. In J. Abramsky (Ed.), *Reasoning, explanation and proof in school mathematics and their place in the intended curriculum.* London: Qualifications and Curriculum Authority.

Tzur, R. (1999). An integrated study of children's construction of improper fractions and the teacher's role in promoting that learning. *Journal for Research in Mathematics Education*, 30(4), 390–416.

Usiskin, Z. (1987). Resolving the continuing dilemmas in school geometry. In M. M. Lindquist and A. P. Shulte (Eds), *Learning and teaching geometry, K-12* (pp. 17–31). Reston, VA: NCTM.

Usiskin, Z. (1988). Conceptions of school algebra and uses of variables. In A. Coxford and A. Shulte (Eds), *The ideas of algebra, K-12. 1988 Yearbook* (pp. 8–19). Reston, VA: The National Council of Teachers of Mathematics.

Usiskin, Z. and Griffin, J. (2008). *The classification of quadrilaterals: A study in definition.* Charlotte, NC: Information Age Publishing.

Vallecillos, A. (1995). Comprensión de la lógica del contraste de hipótesis en estudiantes universitarios [Understanding of the logic of hypothesis testing amongst university students]. *Recherches en Didactique des Mathematiques*, 15, 53–81.

Vallecillos, A. and Batanero, C. (1996). Conditional probability and the level of significance in tests of hypotheses. In L. Puig and A. Gutiérrez (Eds), *Proceedings of the 20th Conference of the International Group for the Psychology of Mathematics Education* (Vol. 4, pp. 271–378). Valencia, Spain: University of Valencia.

van den Berg, O. (1995). 'Progression in measuring': Some comments. *Research Papers in Education*, 10(2), 171–173.

van den Heuvel-Panhuizen, M. and Buys, K. (Eds) (2008). *Young children learn measurement and geometry. A learning-teaching trajectory with intermediate attainment targets for the lower grades in primary school.* Rotterdam: Sense Publishers.

van Dooren, W., De Bock, D., Depaepe, F., *et al.* (2003). The illusion of linearity: Expanding the evidence towards probabilistic reasoning. *Educational Studies in Mathematics*, 53(2), 113–138.

van Hiele, P. M. (1959/1984). A child's thought and geometry. In D. Fuys, D. Geddes, and R. Tischler (Eds), *English translation of selected writings of Dina van Hiele-Geldof and P. M. van Hiele* (pp. 243–252). Brooklyn: Brooklyn College. (Original document in French in 1959. La pensee de l'enfant et la geometrie, *Bulletin de l' Association des Professeurs de Mathematiques de l'Enseignment Public*, 198, 199–205.)

van Hiele, P. M. (1986). *Structure and insight: A theory of mathematics education.* New York: Academic Press.

Vinner, S. (1991). The role of definitions in the teaching and learning of mathematics. In D. Tall (Ed.). *Advanced mathematical thinking* (pp. 65–81). Dordrecht, The Netherlands: Kluwer.

Vlassis, J. (2002). The balance model: Hindrance or support for the solving of linear equations with one unknown. *Educational Studies in Mathematics*, 49(3), 341–359.

Vosniadou, S. and Ortony, A. (1989). Similarity and analogical reasoning: A synthesis. In S. Vosniadou and A. Ortony (Eds), *Similarity and analogical reasoning* (pp. 1–17). Cambridge: Cambridge University Press.

Warren, E. and Cooper, T. (2007). Generalising the pattern rule for visual growth patterns: Actions that support 8 year olds' thinking. *Educational Studies in Mathematics,* 67(2), 171–185.

Watson, A. (2009a). Paper 7: Modelling, problem-solving and integrating concepts. In *Key understandings in mathematics learning.* London: Nuffield Foundation.

Watson, A. (2009b). Notes from the working group on trigonometry. *Proceedings of British Society for Research in Learning Mathematics,* 29(2). Retrieved on 22.10.2010 from: http://www.bsrlm.org.uk/IPs/ip29-2/index.html.

Watson, A. and Mason, J. (1998). *Questions and prompts for mathematical thinking.* Derby: Association of Teachers of Mathematics.

Watson, J. (2001). Longitudinal development of inferential reasoning by school students. *Educational Studies in Mathematics,* 47(3), 337–372.

Watson, J. M. and Moritz, J. B. (1997). Student analysis of variables in a media context. In B. Phillips (Ed.), *Proceedings of the 8th International Congress of Mathematics Education* (pp. 129–147). Hawthorne, Australia.

Watson, J. M. and Moritz, J. B. (2000). Developing concepts of sampling. *Journal for Research in Mathematics Education,* 31(1), 44–70.

Watson, J. M. and Moritz, J. B. (2003). Fairness of dice: A longitudinal study of students' beliefs and strategies for making judgments. *Journal for Research in Mathematics Education,* 34(4), 270–304.

Watson, J. M. and Kelly, B. A. (2005). Cognition and instruction: Reasoning about bias in sampling. *Mathematics Education Research Journal,* 17(1), 24–57.

Wearne, D. and Kouba, V. L. (2000). Rational numbers. In E. A. Silver and P. A. Kenny (Eds), *Results from the Seventh Mathematics Assessment of the National Assessment of Educational Progress* (pp. 163–191). Reston, VA: National Council of Teachers of Mathematics.

Weber, K. (2005). Students' understanding of trigonometric functions. *Mathematics Education Research Journal,* 17(3), 91–112.

Weber, K. (2008). Teaching trigonometric functions: Lessons learned from research. *Mathematics Teacher,* 102(2), 144–150.

White, P. and Mitchelmore, M. C. (1996). Conceptual knowledge in introductory calculus. *Journal for Research in Mathematics Education,* 27(1), 79–95.

Whiteley, W. (1999). The decline and rise of geometry in 20th century North America. In J. G. McLoughlin (Ed.), *Canadian Mathematics Education Study Group* (pp. 7–30). St. John's, NF: Memorial University of Newfoundland.

Wild, C. (2006). The concept of distribution. *Statistics Education Research Journal,* 5(2), 10–26.

Wild, C. J. and Pfannkuch, M. (1999). Statistical thinking in empirical enquiry. *International Statistical Review,* 67(3), 223–265.

Wild, C. J., Pfannkuch, M., Regan, M., and Horton, N. J. (2011). Towards more accessible conceptions of statistical inference. *Journal of the Royal Statistical Society Series A*, 174, Part 2, 1–23.

Williams, J. and Ryan, J. (2000). *Children's mathematics 4–15: Learning from errors and misconceptions*. Maidenhead: Open University Press.

Williams, S. (1991). Models of limit held by college calculus students. *Journal of Research in Mathematics Education*, 22(3), 219–236.

Willson, W. W. (1977). *The mathematics curriculum: Geometry*. Blackie: Glasgow.

Wilson, K., Ainley, J., and Bills, L. (2005). Spreadsheets, pedagogic strategies and the evolution of meaning for variable. In H. L. Chick and J. L. Vincent (Eds), *Proceedings of the 29th annual conference of the International Group for the Psychology of Mathematics Education* (Vol. 4, pp. 321–328). Melbourne, Australia.

Wollman, W. (1983). Determining the sources of error in a translation from sentence to equation. *Journal for Research in Mathematics Education*, 14(3), 169–181.

Yerushalmy, M. (1991). Students' perceptions of aspects of algebraic function using multiple representation software. *Journal of Computer Assisted Learning*, 7(1), 42–57.

Yerushalmy, M. (1997). Designing representations: Reasoning about functions of two variables. *Journal for Research in Mathematics Education*, 28(4), 431–466.

Yerushalmy, M. (2001). Problem solving strategies and mathematical resources: A longitudinal view on problem solving in a function based approach to algebra. *Educational Studies in Mathematics*, 43(2), 125–147.

Yerushalmy, M. (2005). Challenging known transitions: Learning and teaching algebra with technology. *For the Learning of Mathematics*, 25(3), 37–42.

Yerushalmy, M. and Swidan, O. (2012). Signifying the accumulation graph in a dynamic and multi-representation environment. *Educational Studies of Mathematics* [Online First].

Yerushalmy, M., Katriel, H., and Shternberg, B. (2002). *Visual math: The function web book*. Ramat Aviv, Israel: CET. www.cet.ac.il/math/function/english.

Zaslavsky, O., Sela, H., and Leron, U. (2002). Being sloppy about slope: The effect of changing the scale. *Educational Studies in Mathematics*, 49(1), 119–140.

Zazkis, R., Liljedahl, P., and Gadowsky, K. (2003). Conceptions of function translation: Obstacles, intuitions, and rerouting. *Journal of Mathematical Behavior*, 22, 437–450.

Zeiffler, A., Garfield, J., delMas, B., and Reading, C. (2008). A framework to support research on informal inferential reasoning. *Statistics Education Research Journal*, 7(2), 40–58.

Zendrera, N. (2010). Human sciences students' difficulties in parametric tests: A contribution to statistics education. *Proceedings of the Eighth International Conference on the Teaching of Statistics*. Retrieved on 10.2.2012 from: http://www.stat.auckland.ac.nz/~iase/publications/icots8/ICOTS8_7C2_ZENDRERA.pdf [accessed 4 August 2012].

# INDEX